Biochemistry and Behavior

edited by
Solomon H. Snyder

The MIT Press
Massachusetts Institute of Technology
Cambridge, Massachusetts

BIOCHEMISTRY AND BEHAVIOR

Introduction

SOLOMON H. SNYDER

"BIOCHEMISTRY AND BEHAVIOR" is a global designation, which can mean different things to different people. Faced with a choice between examining a single system in depth or providing an overview of several areas, we have selected something of a compromise.

It has long been presumed that endocrine secretions affect, in some vague fashion, mentation and emotion. David de Wied illustrates ways in which it is possible to explore the influence of hormonal secretions upon animal behavior. These approaches are exemplified in his own research on the role of the adrenal cortical hormones and pituitary peptides in mediating specific animal behaviors.

Quite recently the scientific world has become cognizant of the prominent place nutrition or, better, malnutrition occupies in mental function. Richard Wurtman reviews the ways whereby altered nutritional status can influence the brain. As a case in point, he describes efforts in his own laboratory to clarify the effects of circulating amino acids, especially tryptophan, upon the dynamics of brain serotonin.

The biochemistry of memory has been perhaps the most active area of "behavioral chemistry" research. Unfortunately, the variable caliber of many studies has given rise as much to confusion as to new knowledge. Edward Glassman evaluates critically a wide range of investigations of macromolecules and behavior, especially memory mechanisms. Adrian Dunn then describes some thoughtfully conceived and carefully executed studies conducted in the laboratories of Dunn and Glassman, which explicate several difficulties in macromolecule-memory research.

Following this broad panorama of approaches to the biochemistry-and-behavior interface, succeeding chapters focus on a relatively circumscribed area—the relationship of catecholamines to behavior in animals and man. Urban Ungerstedt reviews his epochal behavioral studies with specific lesions of discrete catecholamine tracts in rats produced by stereotaxically implanted 6-hydroxydopamine. Susan Iversen, using the 6-hydroxy-dopamine technique, asks which animal behaviors are mediated by the classic nigrostriatal dopamine pathway?

Sarah Leibowitz has taken a different approach, injecting catecholamines into various sites in the brain and monitoring the resultant behavior, especially eating and drinking.

Of course, ultimately, one would like to apply all this information to man, especially to mental illness. My chapter reviews data bearing upon a possible role of catecholamines in mediating drug effects in schizophrenia. Issues examined critically include: (a) whether phenothiazine drugs exert a selective antischizophrenic action, (b) which drug psychoses provide the best "model schizophrenia" and (c) detailed interactions of antischizophrenic and schizophrenomimetic drugs with catecholamines. Steven Matthysse presents a speculative discussion attempting to integrate what is known of the part played by various brain regions in motor control and feature extraction with brain dopamine and with schizophrenia.

56 Pituitary-Adrenal System Hormones and Behavior

DAVID DE WIED

You will only wear yourself out and wear out all the people who are here.
The task is too heavy for you, you cannot do it by yourself.
Exodus 18:18
New English Bible

Robert Ader, Bela Bohus, Paul Garrud, Willem Hendrik Gispen, Henk Greven, Alan King, Saul Lande, Peter Schotman, Dirk Versteeg, Ivan Urban, Tjeerd van Wimersma Greidanus, Jan Weijnen, and Albert Witter.

ABSTRACT The role of the pituitary-adrenal system and pituitary-adrenal system hormones on acquisition and maintenance of conditioned behavior is the primary subject. Several topics are discussed, including: (a) pituitary-adrenal activity and conditioned behavior; (b) influence of pituitary-adrenal ablation and substitution on acquisition and maintenance of conditioned behavior; (c) structure-activity studies with pituitary-adrenal system hormones; (d) site of the behavioral action of pituitary-adrenal system hormones in the brain, and (e) mode of action of ACTH and ACTH analogues in the brain.

Introduction

THE PITUITARY-ADRENAL system plays an essential role in the defense of the organism against noxious stimuli. Stress rapidly stimulates the release of adrenocorticotrophic hormone (ACTH) from the adenohypophysis. This pituitary hormone activates the adrenal cortex resulting in an increased secretion of glucocorticosteroids. Pituitary ACTH-release is mediated by a corticotrophin-releasing factor or hormone (CRF) of hypothalamic origin, and modulated by various subcortical structures. The function of these structures, in turn, is affected by the level of circulating pituitary-adrenal hormones.

Originally, the activation of the pituitary-adrenal system was regarded mainly to be associated with severe tissue destruction. Only when the assay of hormones produced by the pituitary and adrenal gland became more sophisticated was it found that neurogenic and also psychic stimuli are as potent as tissue damage in stimulating the pituitary-adrenal system.

DAVID de WIED Rudolf Magnus Institute for Pharmacology, Medical Faculty, University of Utrecht, Utrecht, The Netherlands

Evidence for an influence of ACTH and adrenal steroids on the brain was obtained from animal experiments on brain excitability (Woodbury, 1954) and also from clinical observations after these hormones were introduced into the clinic. These investigations suggested that the central nervous system (CNS) contains areas that are sensitive to ACTH and corticosteroids. The influence of pituitary-adrenal system hormones on the brain may therefore be reflected in behavior associated with stress and adaptation.

Behavioral effects of pituitary-adrenal ablation and substitution therapy

ADRENALECTOMY Adrenalectomy does not impair the rat's ability to acquire an avoidance response (Moyer, 1958; Applezweig and Moeller, 1959; Bohus and Endröczi, 1965; de Wied, 1967; van Delft, 1970). In fact, Beatty et al. (1970) found that adrenalectomy attenuated the deleterious effects of high shock in a shuttle box avoidance situation. Weiss et al. (1970) observed a superior acquisition of both active and passive avoidance behavior in adrenalectomized rats. Since the level of circulating ACTH is high in adrenalectomized rats (Hodges and Vernikos-Danellis, 1962), it has been suggested that the ameliorating effect of adrenalectomy is the result of the influence of supraphysiological amounts of endogenous ACTH. Treatment with a glucocorticosteroid that normalized the level of circulating ACTH in adrenalectomized rats, also normalized the avoidance response of these animals (Weiss et al., 1970). However, van Delft (1970) was unable to demonstrate superior avoidance performance in adrenalectomized rats in a pole-jumping avoidance situation. In addition, the administration of glucocorticosteroids did not reduce the performance of the adrenalectomized rats. On the contrary, these steroids tended to stimulate the rate of acquisition of the avoidance response. The U-shaped relationship between performance of an incompletely learned shuttle box avoidance response and the interval between initial and subsequent training trials (Kamin, 1963) has been related to alteration in pituitary-adrenal activity. The poorest performance was found to be accompanied by low corticosterone levels (Brush and Levine, 1966). However, adrenalectomy failed to alter the U-shaped avoidance function (Barrett et al., 1971). The above findings indicate that adrenalectomy per se is associated with normal responding in conditioned avoidance situations and tends under certain conditions to enhance avoidance performance.

HYPOPHYSECTOMY Removal of the pituitary seriously interferes with acquisition of conditioned avoidance behavior. Adenohypophysectomy (de Wied, 1964) or ablation of the whole pituitary (Applezweig and Baudry, 1955; de Wied, 1968) markedly reduced the ability of the rat to acquire a shuttle box avoidance response (Figure 1).

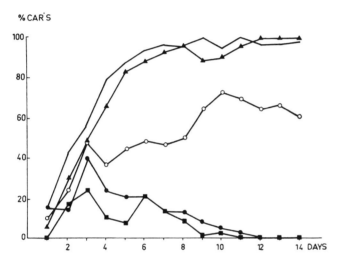

% CAR'S

FIGURE 1 Effect of testosterone propionate, O———O (0.2 mg per 2 days s.c.); dexamethasone phosphate, ■———■ (10 μg per day s.c.); ACTH$_{4-10}$ zinc phosphate, ▲———▲ (20 μg per 2 days s.c.); and placebo (0.5 ml zinc phosphate complex s.c.), ●———● on the rate of acquisition of a shuttle-box avoidance response in hypophysectomized rats, as compared to sham-operated rats, ——— . Hypophysectomy was performed by the transauricular route under ether anesthesia. Animals were allowed to recover from the operation for 1 week after which treatment and conditioning were begun. Ten trials a day were given for 14 days with a variable intertrial interval averaging 60 sec. The sound of a buzzer served as the CS (de Wied, 1967).

Hypophysectomy results in metabolic derangements and physical weakness of the organism; treatment of adenohypophysectomized rats with a substitution therapy, consisting of thyroxin, cortisone, and testosterone, improved the health and condition of these animals and at the same time improved avoidance acquisition (de Wied, 1964). Treatment of hypophysectomized rats with testosterone alone (Figure 1) markedly abolished loss of body weight that normally occurs following removal of the

pituitary gland and facilitated the rate of avoidance learning. Similarly, growth-hormone administration prevented body weight loss and stimulated the rate of acquisition of a shuttle box avoidance response in hypophysectomized rats in a dose dependent manner (de Wied, 1969). Thus, improving the health and condition of the hypophysectomized organism ameliorates behavioral performance.

Treatment of hypophysectomized rats with adrenal maintenance doses of ACTH restored the rate of acquisition of the avoidance response of adenohypophysectomized or hypophysectomized rats toward nearly normal levels (Applezweig and Moeller, 1959; de Wied, 1964, 1968). Lack of adrenal gland hormones cannot account for the behavioral deficiency of hypophysectomized rats. Adrenalectomy, as we have seen, does not impair avoidance acquisition. Moreover, treatment of hypophysectomized rats with a potent glucocorticosteroid, dexamethasone (Figure 1), failed to improve avoidance acquisition in hypophysectomized rats (de Wied, 1971). Accordingly, the influence of ACTH on avoidance acquisition of hypophysectomized rats is probably due to an extra target effect of this hormone, presumably located in the CNS. This was demonstrated by the use of peptides structurally related to ACTH but devoid of corticotrophic activities. Administration of α-MSH, which contains the sequence 1–13 of the ACTH molecule, or even smaller analogues like ACTH$_{1-10}$ or ACTH$_{4-10}$ (Figure 1) restored the rate of acquisition of a shuttle-box avoidance response to nearly the level of sham-operated control animals (de Wied, 1969). Further analysis of the influence of the heptapeptide ACTH$_{4-10}$ (de Wied, 1967) revealed that it does not materially affect endocrine or metabolic functions nor the general health and condition of the hypophysectomized rat as determined by loss of body weight, adrenal and testes weight, blood glucose, and insulin and FFA content of plasma. Motor capacities of these animals were only slightly affected by the administration of ACTH$_{4-10}$ (de Wied, 1968). Interestingly, the sensory capacities of hypophysectomized animals are different from those of sham-operated controls. The shock threshold is much lower in the former animals, and this increased responsiveness to electric shock is not restored to normal by treatment with ACTH$_{1-10}$ (Gispen et al., 1970a).

Hypophysectomy attenuates not only active but also passive avoidance behavior (Anderson et al., 1968; Weiss et al., 1970; Lissák and Bohus, 1972). Treatment of hypophysectomized rats with ACTH improves avoidance responding in both situations, supporting the hypothesis that the behavioral deficiency of hypophysectomized rats is not due to motor activity disturbances.

Table I

Effect of chain length shortening of $ACTH_{1-10}$ at the amino end on extinction of a pole-jumping avoidance response

	1 2 3 4 5 6 7 8 9 10	Effect*
$ACTH_{1-10}$	H-Ser-Tyr-Ser-Met-Glu-His-Phe-Arg-Trp-Gly-OH	100
$ACTH_{2-10}$	H-Tyr————————————————	100
$ACTH_{3-10}$	H-Ser————————————	100
$ACTH_{4-10}$	H-Met————————————	100
$ACTH_{5-10}$	H-Glu————————	50
$ACTH_{6-10}$	H-His————————	0
$ACTH_{7-10}$	H-Phe————————	0

*Approximated potency.

Pituitary-adrenal system hormones and their effects on acquisition and extinction of conditioned behavior

ACTH AND ACTH ANALOGUES   Pituitary-adrenal system hormones are not particularly effective in changing acquisition of active avoidance behavior. Daily administration of ACTH did not alter the rate of acquisition of a shuttle-box avoidance response of intact (Murphy and Miller, 1955) or posterior lobectomized rats (de Wied, 1969). Apparently, removal of endogenous ACTH by hypophysectomy is necessary to demonstrate an effect of this polypeptide on avoidance learning. However, Guth et al. (1971) demonstrated recently that injection of ACTH during acquisition of an appetitive response significantly increased lever press responses late in the training period. This effect was more marked following stringent control of environmental stimulation.

Extinction of an avoidance response is much more sensitive to the treatment with ACTH. Murphy and Miller (1955) found that ACTH administered during shuttle box training resulted in increased resistance to extinction. A more pronounced effect on extinction, however, is found when ACTH is administered during the extinction period (de Wied, 1967). Although the long-term administration of ACTH is accompanied by hypercorticism, the influence of ACTH is independent of its action on the adrenal cortex. ACTH is also active in adrenalectomized rats (Miller and Ogawa, 1962). Moreover, synthetic α-MSH, purified β-MSH and also the smaller ACTH analogues like $ACTH_{1-10}$ and $ACTH_{4-10}$ (de Wied, 1966; de Wied and Bohus, 1966) increase resistance to extinction of a shuttle box as well as a pole-jumping avoidance response. Interestingly, $ACTH_{11-24}$ is ineffective in this respect (de Wied et al., 1968). Thus, the influence of ACTH on the maintenance of aversively motivated behavior is located in the amino end of the molecule, presumably in the heptapeptide $ACTH_{4-10}$, a sequence that is common to ACTH, α- and β-MSH. This conclusion is in accord with observations in dogs in which intracysternal administration of ACTH

and also of α-, β-MSH and $ACTH_{4-10}$ caused a stretching syndrome. Further shortening of the peptide $ACTH_{4-10}$ from the amino end (Greven and de Wied, 1967) interfered with the potency of the inhibitory effect of the peptide on extinction of a pole-jumping response (Table I). Shortening from the carboxyl end (Greven and de Wied, 1973) remained without marked consequences for the behavioral effects until the amino acid phenylalanine at position 7 was removed (Table II). It was surprising that the behavioral activity depended on such a small peptide. Interestingly, this also holds for MSH, since the minimal requirement for a residual activity on expansion of melanophores for α-MSH is the sequence 6–9 (Otsuka and Inouye, 1964). (MSH: melanocyte-stimulating hormone.)

Table II

Effect of chain length shortening of $ACTH_{4-10}$ at the carboxyl end on extinction of a pole-jumping avoidance response

		Effect*
$ACTH_{4-10}$	H-Met-Glu-His-Phe-Arg-Trp-Gly-OH	100
$ACTH_{4-9}$	————————————OH	100
$ACTH_{4-8}$	——————————OH	100
$ACTH_{4-7}$	————————OH	30–100
$ACTH_{4-6}$	——————OH	10
$ACTH_{4-5}$	————OH	10
$ACTH_{5-6}$	———OH	< 10

*Approximated potency.

ACTH affects passive avoidance behavior of intact rats as well. Lissák et al. (1957) were the first to show an effect of ACTH in a passive avoidance situation, and subsequently Levine and Jones (1965) found that ACTH affected acquisition of a passive avoidance response. Using a simple step-through passive avoidance procedure, we found that ACTH administration 1 hr prior to the first retention trial markedly increased avoidance latencies. The same occurred when $ACTH_{1-10}$ was given in contrast to $ACTH_{11-24}$ or $ACTH_{25-39}$ (Table III). Thus, also in this test, the behaviorally active core of ACTH appeared to reside in the amino end of the molecule.

TABLE III

Effect of sequences of the ACTH molecule on avoidance latency in a step-through passive avoidance test

	Median Preshock* Latency Mean of 3 Trials (sec)	Median Postshock Latency (sec)	
		Trial 1	Trial 2
Saline 0.5 ml s.c.	2	10	2
ACTH$_{1-10}$ 10 μg s.c.	3	175†	6
ACTH$_{1-10}$ 30 μg s.c.	3	225†	7
ACTH$_{11-24}$ 10 μg s.c.	2	12	3
ACTH$_{11-24}$ 30 μg s.c.	2	78†	4
ACTH$_{25-39}$ 10 μg s.c.	2	8	3
ACTH$_{25-39}$ 30 μg s.c.	3	49†	3

6 animals per group.
*(0.25 mA; 1 sec).
†Significantly different from saline treated controls.

A number of recently performed studies indicate that ACTH and related peptides not only affect extinction of "fear" motivated behavior. ACTH and also MSH inhibits extinction of an appetitive response as well (Sandman et al., 1969; Leonard, 1969; Gray, 1971; Guth et al., 1971), and the same holds for ACTH$_{4-10}$ (Garrud and de Wied, 1972 (unpublished observations)). ACTH has been shown to suppress aggression in intact and adrenalectomized mice (Brain, 1971; Pasley and Christian, 1972). In various behavioral situations, however, ACTH or related peptides have been reported to be inactive. ACTH$_{1-10}$ did not, or only slightly, affect ambulation, rearing, grooming, or the production of fecal boli in an open field (Bohus and de Wied, 1966; Weijnen and Slangen, 1970; Hadžović and de Wied, 1971). Escape behavior in a runway or the responsiveness to electric shock was also not affected by chronic administration of ACTH$_{1-10}$ (Bohus and de Wied, 1966; Gispen et al., 1970a). Negative results that were obtained in other test situations were reported by Weijnen and Slangen (1970). Accordingly, except for a stretching crisis in dogs (Ferrari et al., 1963) and an effect on aggressive behavior in mice (Brain, 1971; Pasley and Christian, 1972), ACTH and related peptides seem to be effective mainly in modifying conditioned behavior.

ACTH ANALOGUES WITH A D-ISOMER PHENYLALANINE AT POSITION 7   One of the first analogues of ACTH that was synthetized was the decapeptide ACTH$_{1-10}$ in which the phenylalanine residue in position 7 was replaced by the D-isomer. This peptide was found to facilitate extinction of the avoidance response in the shuttle box (Bohus and de Wied, 1966). This result suggested that the D-form peptide antagonized the action of normally

occurring ACTH and related peptides which inhibit extinction of avoidance behavior. For this reason the influence of ACTH$_{1-10}$ (7-D-phe) was studied in the absence of ACTH and MSH in hypophysectomized rats. The D-form peptide facilitated the rate of extinction of a shuttle box avoidance response in hypophysectomized rats as well. The effect was even stronger than in sham-operated controls. Subsequently, it was found that ACTH$_{4-10}$ (7-D-phe) (Figure 2) had an effect similar to that of the D-isomer decapeptide (de Wied and Greven, 1968). Preliminary observations indicate that amino acids 10, 9, and 8 can be removed from the carboxyl end of the (7-D-phe) peptides as in the all -L- peptides, without causing severe damage to the potency of the D-form peptide in facilitating extinction of a pole-jumping avoidance response.

% CAR'S

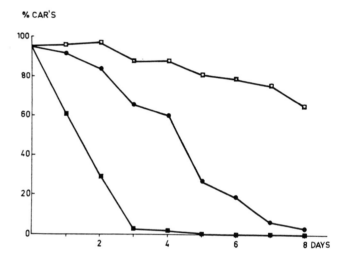

FIGURE 2   Effect of ACTH$_{4-10}$ (7-L-phe) zinc phosphate, □———□ (10 μg per 2 days s.c.); and ACTH$_{4-10}$ (7-D-phe) zinc phosphate, ■———■ (10 μg per 2 days s.c.) on the rate of extinction of a shuttle box avoidance response as compared to placebo treatment, ●———● . Male rats were conditioned in the shuttle box. Ten trials a day were given with a variable intertrial interval of 60 sec. The sound of a buzzer served as the CS. Animals were trained till they achieved the conditioning criterion of 8 or more avoidances out of 10 for 3 consecutive days. Immediately thereafter extinction trials were run with the same schedule as in acquisition except that the US of shock did follow the CS if the animal failed to respond.

The D-form peptides facilitate extinction of a shuttle box and a pole-jumping avoidance response. They also seem to facilitate extinction of an appetitive response (Garrud and de Wied, 1972(unpublished observations)). However, their effect on passive avoidance behavior is not opposite to that of all-L-ACTH analogues as in active avoidance behavior (Table IV). In a simple step-through passive avoidance test (Ader et al., 1972), these peptides increased latency to enter a dark box like all-L-peptides.

TABLE IV

Effect of various D-isomer ACTH analogues on avoidance latency in a step-through passive avoidance response

Shock Intensity	Treatment	Median Preshock* Latency Mean of 3 trials (sec)	Median Postshock Latency (sec)	
			Trial 1	Trial 2
0.0 mA	Saline 0.5 ml s.c.	4	2	2
	$ACTH_{1-10}$ (7-D-phe) 10 μg s.c.	2	3	3
0.25 mA	Saline 0.5 ml s.c.	3	10	4
	$ACTH_{1-10}$ (7-D-phe) 10 μg s.c.	4	74†	43
	Saline 0.5 ml s.c.	2	22	10
	8-Lys-$ACTH_{4-9}$ (7-D-phe) 30 μg s.c.	3	250†	300†
0.50 mA	$ACTH_{4-10}$ 30 μg s.c.	2	300†	45†
	Saline 0.5 ml s.c.	4	87	59
	$ACTH_{1-10}$ (7-D-phe) 10 μg s.c.	4	276†	201†

5 animals per group.

*1 sec.

†Significantly different from saline treated controls.

Interestingly, the influence of the D-isomer peptides lasted longer than that of the all-L-analogues. The influence of the (7-D-phe) peptides in the passive avoidance test support our concept that these peptides do not influence conditioned behavior by competing with all-L-peptides for the same receptor, but possibly affect other nervous structures, resulting in an opposite effect on extinction of active avoidance behavior.

D-ISOMER SUBSTITUTION IN OTHER POSITIONS Subsequently it was investigated whether the reversal of active avoidance behavior by D-isomer substitution is restricted to the phenylalanine at position 7. Therefore, successive replacement of each of the amino acid residues was performed in the hexapeptide 8-lys-$ACTH_{4-9}$, which was found to be as active as $ACTH_{4-10}$. It is of interest to note that substitution of arginine by lysine, as in 8-lys-$ACTH_{4-9}$, is accompanied by loss of steroidogenic activity and of MSH activity (Chung and Li, 1967),while the behavioral activity remains unaltered in the peptides 8-lys-$ACTH_{1-10}$ (Greven and de Wied, 1967) and 8-lys-$ACTH_{4-9}$.

The amino acid sequences of the substituted hexapeptides are shown in Table V. None of the replacements

induced facilitation of extinction of a pole-jumping avoidance response. In contrast, these peptides invariably delayed extinction of the avoidance response as normally found with all-L-peptides derived from ACTH. In most cases replacement by a D-antipode potentiated the inhibitory effect of all-L-8-lys-$ACTH_{4-9}$. This potentiation was most marked with the lysine residue at position 8 in the D-configuration. The observed potentiation may be explained by enhanced resistance to proteolytic breakdown of the D-isomer peptides. These results clearly show that position 7 is specific for facilitation of extinction of an active avoidance response and add further evidence for the specificity of the influence of ACTH analogues on conditioned behavior.

A HIGHLY POTENT NEWLY SYNTHETIZED ACTH ANALOGUE Replacement of the lysine at position 8 by a D-isomer potentiated the inhibitory effect on extinction of the avoidance response more than 10 times. Substitution of tryptophan by phenylalanine in position 9 in the 8-D-lys-$ACTH_{4-9}$-peptide elicited a more than hundredfold

TABLE V

Amino acid sequence of 8-lys-$ACTH_{4-9}$ with various D-amino acid substitutions

H-D-Met- Glu - His -Phe- Lys - Trp-OH
H-Met -D-Glu- His -Phe- Lys - Trp-OH
H-Met - Glu -D-His-Phe- Lys - Trp-OH
H-Met - Glu - His -Phe-D-Lys- Trp-OH
H-Met - Glu - His -Phe- Lys -D-Trp-OH

TABLE VI

Effect of amino acid substitution in 8-lys-$ACTH_{4-9}$ on extinction of a pole-jumping avoidance response

4	5	6	7	8	9

H-Met-Glu-His-Phe-Lys-Trp-OH
0 ↑

H-Met-Glu-His-Phe-Lys-Trp-OH
H-Met-Glu-His-Phe-D-Lys-Trp-OH
H-Met-Glu-His-Phe-D-Lys-Phe-OH
0 ↑

H-Met-Glu-His-Phe-D-Lys-Phe-OH

potentiation of the behavioral effect (Table VI). Oxidation of the methionine to the sulfoxide evoked a three- to tenfold increase in behavioral potency. Interestingly, these modifications cause a marked reduction in the activity of ACTH in the adrenal cortex (Hofmann et al., 1970; Dedman et al., 1955). This accentuates again the different structural requirements for corticotrophic and behavioral activity. The combination of the three potentiating modifications was undertaken and yielded a peptide with a behavioral activity which appeared to be more than a thousandfold stronger than that of $ACTH_{4-10}$. The injection of this new peptide in nanogram quantities causes a marked inhibition of extinction of a pole-jumping avoidance response.

CORTICOSTEROIDS Administration of glucocorticosteroids does not modify active avoidance learning. In most studies, dexamethasone was used, which is extremely potent in blocking ACTH-release. Administration of dexamethasone did not significantly alter the somewhat decremented avoidance acquisition of adrenal demedullated rats in a shuttle box (Conner and Levine, 1969). In adrenalectomized rats, however, dexamethasone tended to stimulate avoidance acquisition of rats in a pole-jumping avoidance test (van Delft, 1970). In contrast, Weiss et al. (1970) found that glucocorticosteroid treatment normalized avoidance responding of adrenalectomized rats which performed superior in both active and passive avoidance behavior. Latencies to resume licking after a single grid shock were unaffected by injection of dexamethasone or corticosterone (Pappas and Gray, 1971). Poor performance of an incompletely learned shuttle box avoidance response that depended on the interval between original training and subsequent training sessions was accompanied by low corticosterone levels in the circulation (Brush and Levine, 1966). Increasing the glucocorticoid level in the blood by administration of ACTH or hydrocortisone induced high responding under these conditions (Levine and Brush, 1967). However, Kasper-Pandi et al. (1970) failed to affect performance with extremely high amounts of dexamethasone under these conditions. Interestingly, a greater stability in interresponse time is induced in free operant avoidance behavior in rats after the administration of dexamethasone (Wertheim et al., 1967). It is possible therefore that glucocorticosteroids influence the efficiency with which animals perform an active avoidance task.

In contrast to affecting acquisition of active avoidance behavior, corticosteroids seem to be more effective in modifying passive avoidance behavior. Administration of a single dose of cortisone, 3 hr prior to shock, led to a suppression of passive avoidance in a light-dark avoidance situation. The effect of corticosteroids appeared to be a function of shock intensity. The higher the intensity, the more corticosteroids were needed to suppress passive avoidance behavior (Bohus et al., 1970). Corticosteroids also suppressed passive avoidance behavior motivated by fear or thirst (Bohus, 1971).

Chronic treatment with rather high doses of long acting ACTH during extinction, as was used in our earlier experiments (de Wied, 1967), invariably caused adrenal hypertrophy and a marked increase in the level of circulating corticosterone. Since at that time an extra target effect of ACTH on behavior was not generally accepted and most studies were performed with the whole ACTH molecule, it was felt necessary to study the influence of corticosteroids on the rate of extinction of conditioned avoidance behavior. Daily administration of moderate amounts of glucocorticosteroids like corticosterone and dexamethasone facilitated extinction of a shuttle box avoidance response. The influence of the mineralocorticosteroid aldosterone seemed to be less pronounced in this respect. This was interpreted to indicate that the behavioral influence of corticosteroids resided in the glucocorticosteroid portion of the molecule (de Wied, 1967). Using a pole-jumping situation, van Wimersma Greidanus (1970) showed, however, that the behavioral effects of adrenal steroids are associated neither with gluco- nor with mineralocorticosteroid activity. Progesterone, 19-norprogesterone and pregnenolone were as potent as corticosterone, while cholesterol, hydroxydione (Viadril), testosterone, and estradiol were ineffective. Common features of behaviorally active steroids appeared to be their double bonds in ring A or B, and their ketogroup of OH-group at C_3 (Figure 3). The ketogroup at C_{20} is important for the potency of the effect, but not essential.

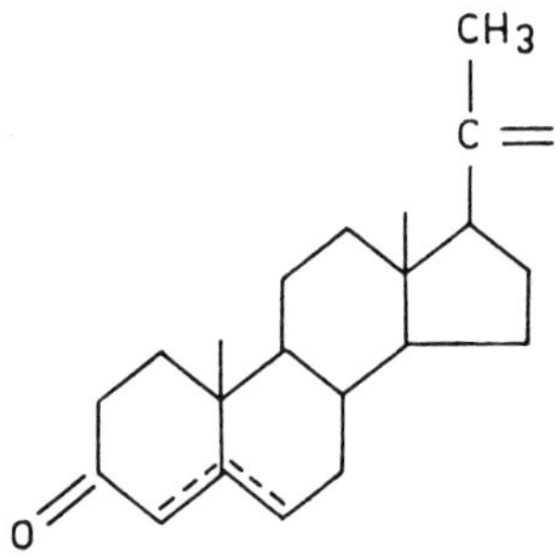

FIGURE 3 Common features of the steroid nucleus necessary to induce facilitation of extinction of a pole-jumping avoidance response.

Since daily administration of glucocorticosteroids inhibits the release of ACTH, one could argue that the influence of these steroids might be explained simply on the basis of a blockade of pituitary-ACTH release. However, the effect of glucocorticosteroids on extinction of a shuttle-box avoidance response did not correlate with the rate of inhibition of ACTH release. In addition, both corticosterone and dexamethasone facilitated extinction of the avoidance response in hypophysectomized rats. Thus, the influence of these steroids on behavior is independent of ACTH and an intrinsic property of the steroid molecule in the CNS. This was supported by experiments in which cortisol was implanted in various areas of the brain (Bohus, 1968). Implantation of cortisol in the median eminence, which effectively inhibited the release of ACTH, had a modest effect on the rate of extinction of a shuttle box avoidance response. The effect was stronger the more ACTH release was suppressed. However, implantation of cortisol in the mesencephalic reticular formation markedly facilitated extinction but hardly reduced ACTH release. Corticosteroids therefore may have a dual effect on extinction of avoidance behavior, one through inhibition of ACTH release and one probably more important through a direct action on the CNS.

Site of action of pituitary-adrenal system hormones in the brain

The site of action of pituitary-adrenal system hormones in the brain has been explored in rats bearing lesions in the thalamic region and by implantation of ACTH analogues and various steroids in the brain. Lesions were made in the thalamic region because this area has been implicated in acquisition and extinction of conditioned avoidance behavior (Vanderwolf, 1964; Thompson, 1963; Rich and Thompson, 1965; Cardo, 1965; Delacour et al., 1966; Delacour, 1970). Rather extensive lesions in the midline thalamic reticular area produced severe deficits in the rate of acquisition of a shuttle-box avoidance response and in escape behavior. Smaller lesions in this region interfered with avoidance acquisition but not with escape behavior (Bohus and de Wied, 1967a). Bilateral destruction of the nucleus parafascicularis did not materially affect avoidance learning but facilitated extinction of the avoidance response. In rats bearing lesions in the nucleus parafascicularis, α-MSH was unable to affect extinction in amounts in which it causes resistance to extinction in intact rats (Bohus and de Wied, 1967b). These results suggested that the nucleus parafascicularis was implicated in the behavioral effect of ACTH analogues.

To localize the action of ACTH analogues and steroids in the brain more specifically, implantation of these compounds in various subcortical structures was undertaken (van Wimersma Greidanus and de Wied, 1969, 1971).

$ACTH_{1-10}$ and $ACTH_{1-10}$ (7-D-phe) were used as the ACTH analogues and corticosterone and dexamethasone (21-Na-phosphate) as the steroids. ACTH analogues had a behavioral effect, i.e., delay with $ACTH_{1-10}$ or facilitation with $ACTH_{1-10}$ (7-D-phe) of extinction of a pole-jumping avoidance response, similar to that of systemic administration, when implanted into the region of the rostral mesencephalon and the caudal diencephalon at the posterior thalamic level or in the cerebrospinal fluid. Ineffective sites were the nucleus ventralis thalami, the nucleus anterior medialis thalami, the nucleus reuniens, the globus pallidus, the nucleus accumbens, the fornix, and the hippocampus.

Implantation of dexamethasone facilitated the rate of extinction of a pole-jumping avoidance response when implanted into various areas in the median and posterior thalamus and in the ventriculus lateralis. Corticosterone also facilitated extinction, but mainly if implanted in or near the nucleus parafascicularis. Implantation of these two steroids in other areas like hippocampus, nucleus septi lateralis, nucleus caudatus putamen, nucleus interstitialis terminalis, and nucleus ventralis thalami did not result in a modification of extinction. However, Bohus (1968) showed that cortisone implantation in the anterior hypothalamus, septum, amygdala, or dorsal hippocampus facilitated extinction of a shuttle-box avoidance response. Figure 4 gives a schematic representation of the brain areas that were found sensitive to peptides and steroids. Interestingly, the two ACTH analogues more or less act in the same structures in the brain. They affect areas in the ascending reticular system, mainly in the rostral mesencephalic caudal diencephalon area at the thalamic level; in particular in the posterior thalamus and more specifically in the nuclei parafascicularis. The steroids act, in addition to the ascending reticular system, in limbic forebrain regions. The opposite action of ACTH and corticosteroids seems to be located in the thalamic area. Facilitation of extinction, as induced by implantation of cortisone in that area, was counteracted by systemic administration of ACTH (Bohus, 1970).

Cardo (1965, 1967) has shown that the nonspecific thalamic nuclei like the nucleus parafascicularis and centrum medianum play an important role in the maintenance of conditioned avoidance behavior. Lesions in these areas interfere with avoidance performance, while electrical stimulation leads to resistance of extinction. Delacour (1970) suggested that the parafascicular-centrum - medianum complex is involved in the interaction between defensive motivation and some mechanisms of avoidance responding in certain avoidance situations. However, a number of complex food-reinforced tasks also are affected by lesions of this complex. The fact

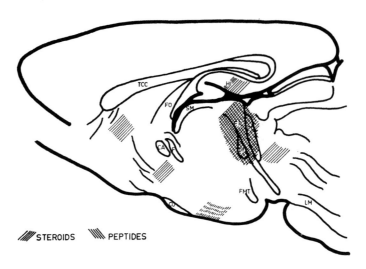

STEROIDS   PETIDES

FIGURE 4 Schematic representation of sites in the brain sensitive to the behavioral effect of ACTH analogues ($ACTH_{1-10}$ [7-L-phe and 7-D-phe]) and corticosteroids. *Abbreviations:* TCC, truncus corporis callosum; Hi, hippocampus; FO, fornix; SM, stria medullaris thalami; FR, fasciculus retroflexus; pf, nucleus parafascicularis; CA, commissura anterior; F, columna fornicis; FMT, fasciculus mamillothalamicus; CO, chiasma opticum; LM, lemniscus medialis.

that ACTH (Gray, 1971; Guth et al., 1971) and α-MSH (Sandman et al., 1969) delay extinction of an appetitive response indicates that the role of the nucleus parafascicularis in conditioned behavior may be of a more general character.

Macromolecular effects of ACTH analogues in the brain

As shown above, hypophysectomy interferes with normal avoidance learning in a shuttle box. This deficient behavior of the hypophysectomized rat might be reflected in biochemical alterations in the brain (Versteeg et al., 1972). In addition, the treatment with ACTH analogues, which restored avoidance learning of the hypophysectomized rat, might at the same time reveal specific biochemical changes that could possibly be related to the influence of ACTH peptides in the brain. Recent studies suggest a close relationship between macromolecule processes in the brain and behavior (Hydén and Lange, 1965; Gaito, 1966; Glassman, 1967). In peripheral organs such as liver, hypophysectomy causes changes in RNA and protein metabolism (Korner, 1964; Tata and Williams-Ashman, 1967; Gupta and Talwar, 1968). Such alterations can be detected in the brain as well. A marked decrease in RNA content in the brain occurs as a result of hypophysectomy. This effect is confined to brainstem areas and the frontal part of the cortex (de Vellis and Inglish, 1968; Gispen et al., 1972). The reduction in RNA content is localized in the microsomal fraction as reflected in a decrease in the number of large polysomes (Gispen and Schotman, 1970). These changes

are accompanied by a decrease in incorporation of radioactive uridine into rapidly labeled RNA (Gispen et al., 1970b) and of radioactive leucine into proteins of all subcellular fractions (Schotman et al., 1972). From these experiments it follows that RNA synthesis in the brainstem of hypophysectomized rats is decreased. The same can be concluded for protein synthesis from the decrease in polyribosomes and in the labeling of proteins. Because protein content of the brain of hypophysectomized rats is not different from that of sham-operated controls, the deficient protein synthesis is restricted to a small fraction of rapidly turning over proteins.

A correlation between acquisition of learned behavior and the synthesis of rapidly labeled RNA (Zemp et al., 1966) and rapidly turning over proteins (Hydén and Lange, 1968) in the brain has been reported. The reduction of these processes in the brain of hypophysectomized rats might therefore be related to the deficient behavior of these animals. Treatment of hypophysectomized rats with $ACTH_{1-10}$ for 14 days restored performance of a shuttle-box avoidance response and at the same time normalized the optical density pattern of brain stem polysomes (Gispen and Schotman, 1970; Gispen et al., 1970b). Interestingly, peptide treatment per se did not restore the polysome pattern of hypophysectomized rats. This only occurred if peptide treatment was combined with avoidance training (Table VII). The primary action of $ACTH_{1-10}$ therefore might be related to other biochemical parameters. It appeared that treatment with $ACTH_{1-10}$ did not affect the incorporation of radioactive uridine into rapidly labeled RNA. However, $ACTH_{1-10}$ treatment did restore the rate of incorporation

TABLE VII

Effect of chronic ACTH$_{1-10}$ treatment, with (I) or without (II) shuttle-box training, on brainstem polysome profiles characterized by absorbance ratios. (From Ph.D. thesis. W. H. Gispen, Utrecht, 1970.)

Experiment	Treatment	Small Polysomes, Monosomes	Large Polysomes, Monosomes	Total Number CARs* out of 100 Trials
I	Hypox + placebo	1.34 ± 0.06†	1.17 ± 0.05‡	23 ± 1 (7)
	Hypox + ACTH$_{1-10}$	1.30 ± 0.05	1.63 ± 0.04‡	75 ± 3 (3)
II	Hypox + placebo	1.32 ± 0.05	1.60 ± 0.11§	— (5)
	Hypox + ACTH$_{1-10}$	1.32 ± 0.07	1.65 ± 0.11§	— (5)

*Conditioned avoidance responses.
†Mean ± standard error of the mean.
‡$p < 0.05$ (modified t-test).
§$p > 0.10$ (modified t-test).
( ) Number of groups of 3 rats.

TABLE VIII

Effect of chronic treatment with various ACTH analogues on incorporation of radioactivity into proteins of brainstem of hypophysectomized rats, 5 min following injection of (4.5 ^{3}H-leucine) (Ph.D. Thesis, P. Schotman, Utrecht, 1971.)

	Percent Total Recovery		
Treatment	Acid Soluble Radioactivity	Acid Insoluble Radioactivity	
ACTH$_{1-10}$ (7-L-phe)	63.2 ± 2.6*	36.8 ± 2.6	(6)
Placebo	71.2 ± 1.1	28.8 ± 1.1	(6)
Difference (%)	−11.2†	+27.8†	
ACTH$_{1-10}$ (7-D-phe)	80.3 ± 2.5	19.7 ± 2.5	(6)
Placebo	72.7 ± 0.8	27.3 ± 0.8	(6)
Difference	+10.5†	−27.8†	
ACTH$_{11-24}$	74.5 ± 3.4	25.5 ± 3.4	(3)
Placebo	72.9 ± 5.1	27.1 ± 3.4	(3)

*Mean ± standard error of the mean.
†Significant difference.
( ) Number of animals.

of radioactive leucine into rapidly turning over proteins towards incorporation rates of intact rats (Table VIII). This is in accord with findings of Reading and Dewar (1971) in intact rats treated with ACTH$_{4-10}$; they found an increased incorporation of radioactive leucine into brain protein 48 hr after injection of the radioactive amino acid. High doses of ACTH itself also stimulated RNA and protein synthesis in the brain of intact rats (Jakoubek et al., 1970) although the interpretation of these results is complicated by a possible action of corticosteroids on brain metabolism (McEwen et al., 1970).

The restoration of the rate of incorporation of radioactive leucine into rapidly turning over proteins of the brainstem of hypophysectomized rats by ACTH$_{1-10}$ seems related to the restoration of the behavioral deficiency of the hypophysectomized rat in the shuttle box.

This is supported by experiments in which ACTH$_{1-10}$ (7-D-phe) was used. This peptide further deteriorates the already deficient avoidance behavior of hypophysectomized rats and at the same time lowers the rate of incorporation of radioactive leucine in the brainstem proteins. Moreover, ACTH$_{11-24}$, which did not affect avoidance acquisition, also failed to alter this parameter of protein synthesis (Table VIII). These results suggest that a disturbance in the synthesis of rapidly turning over proteins of brainstem origin is responsible for the deficient avoidance behavior of hypophysectomized rats. Polysomal aggregation does not seem to be involved in the effect of ACTH$_{1-10}$ per se but probably needs environmental stimulation (Appel et al., 1967) and training (Dellweg et al., 1968), since the effect of ACTH$_{1-10}$ on polysomes was found only in hypophysectomized rats subjected to shuttle box conditioning.

Concluding remarks

Although pituitary-adrenal system hormones affect the formation and maintenance of conditioned behavior, ACTH shares its influence with α- and β-MSH and even smaller analogues of these peptides, which lack corticotrophic activities. The same holds for adrenal steroids. The action of these steroids on conditioned behavior, which is opposite to that of ACTH, is not restricted to those steroids whose production is ACTH dependent. This makes it rather difficult to assign a role to the pituitary-adrenal system per se in conditioned behavior. ACTH and glucocorticosteroids, however, are secreted in high quantities in response to stress, and the system is highly active in many behavioral situations. One can expect therefore that alterations in the level of these hormones affect ongoing behavior associated with stress. It is well known that the pituitary-adrenal system plays an essential role in adaptation (Selye, 1950).

Acquisition and extinction of conditioned behavior indicate a behavioral adaptation to environmental changes. The influence of ACTH and glucocorticosteroids on acquisition and extinction of conditioned behavior should therefore be considered in an adaptive framework.

Various hypotheses have been put forward in the literature to account for the behavioral effects of pituitary-adrenal system hormones. A number of keywords in these hypotheses are anxiety, fear, memory, learning, arousal, motivation, internal inhibition, timing behavior, state dependent learning, etc. (de Wied et al., 1972). Weiss et al. (1970) postulated that ACTH increases excitability which leads to an increase in generalized fear or anxiety in fear situations.

Data obtained from electrophysiological studies indicate that ACTH and analogues have a central excitatory action; corticosteroids may either stimulate or depress activity in the brain (Kawakami et al., 1966; Korányi et al., 1971a, 1971b; Pfaff et al., 1971; Phillips and Dafny, 1971; Sawyer et al., 1968; Steiner, 1970). This suggests that pituitary-adrenal system hormones affect the arousal of certain structures in the CNS. Alteration in the state of arousal of these structures may determine the motivational influence of environmental stimuli. This phenomenon is not only related to avoidance but also to approach behavior, since it has recently been shown that acquisition and extinction of appetitive and drinking responses are also modified by pituitary-adrenal system hormones (Bohus, 1970; Gray, 1971; Gray et al., 1971; Guth et al., 1971; Sandman et al., 1969).

Indications on the locus of action were obtained from lesion and implantation studies. These suggest that the thalamic area and the limbic forebrain are structures sensitive to the behavioral effect of pituitary-adrenal system hormones. The nonspecific thalamic reticular area has an important integrative function (Cardo, 1965; 1967) because all incoming information converges in this structure. Implantation of peptides and steroids in this region effectively alters extinction of conditioned behavior. It is possible therefore that pituitary-adrenal system hormones and related compounds modulate the neural transmission in this nodal point of sensory integration. In this respect it is of interest to note that reciprocal alterations in sensory detection can be demonstrated during circadian alterations of pituitary-adrenal activity (Henkin, 1970). When the levels of glucocorticosteroids in plasma are low, taste detection is high, and vice versa. The marked changes in pituitary-adrenal activity that accompany numerous behavioral situations can therefore be expected to influence integration of extrinsic and intrinsic information, thereby affecting acquisition and maintenance of conditioned behavior.

The behavioral effect of ACTH appears to reside in not more than 4 amino acids. The sequence $ACTH_{4-7}$ (H-met-glu-his-phe-OH) is nearly as active as that of larger sequences like α- and β-MSH. Removal of the amino acid methionine at the amino end and/or phenylalanine at the carboxyl end reduces the behavioral effect. Interestingly, the minimal requirement for a residual activity on melanophore dispersion of α-MSH is the sequence 6–9 (Otsuka and Inouye, 1964). That such a small part of the ACTH molecule suffices for its behavioral activity is reminiscent of similar observations with regard to "tetragastrin," a peptide that exerts essentially the same biological activity as the complete gastrin molecule (Tracy and Gregory, 1964). In view of this it is possible that ACTH and MSH and other peptides act as prohormones from which the active sequence is released by specifically localized enzymes, in the same way as has been suggested by Walter and associates (Celis et al., 1971) for oxytocin that seems to act as a prohormone for the MSH-releasing and the MSH-release-inhibiting factor.

Behaviorally, the most interesting amino acid in the ACTH and MSH molecule appears to be 7-phenylalanine. Replacement of this amino acid by its D-isomer reverses the behavioral effect in active avoidance studies. This reversal is not obtained when neighboring amino acids are replaced by their respective D-isomers. The fact that the 7-D-phenylalanine analogues are effective in the absence of ACTH and MSH and, in addition as recently has been found, influence a passive avoidance response in essentially the same way as the all-L-analogues indicates that L- and D-form analogues act in a nondirect antagonistic manner. However, receptor studies are necessary to determine the specific action of these compounds. Such studies should be performed with radioactive labeled $ACTH_{4-10}$ in the same way as Hofmann et al. (1970) performed with $ACTH_{1-24}$ in the adrenal cortex to determine the active part of this molecule responsible for corticoidogenesis.

A remarkable increase in potency of the behavioral effect of the hexapeptide 8-lys-$ACTH_{4-9}$ was obtained by three modifications in the molecule. Replacement of methionine at position 4 by methionine sulfoxide, of lysine at position 8 by the D-isomer, and of tryptophan at position 9 by phenylalanine produced a peptide, which is active in nanogram quantities. This potentiating effect probably results from a protective influence by the respective substituted amino acids on the active core of the molecule in metabolic degradation.

Studies on the action of ACTH analogues on macromolecule metabolism will probably be very fruitful with respect to mode of action of these peptides. Hypophysectomy, which interferes with the ability to acquire a rather complicated behavioral response in the shuttle box, decreases the rate of incorporation of radioactive leucine into rapidly turning over proteins in the brainstem. The treatment of hypophysectomized rats with $ACTH_{1-10}$ not only restores avoidance acquisition but also the rate of incorporation of radioactive leucine. Moreover, the administration of $ACTH_{1-10}$ (7-D-phe), which deteriorates the already deficient avoidance behavior of hypophysectomized rats, further decreases the incorporation of radioactive leucine of the already reduced incorporation into rapidly turning over proteins. The function of rapidly turning over proteins for the formation of new behavior can be derived from studies with S100-proteins (Hydén and Lange, 1970) and from the amnesic action of antibiotics that interfere with protein metabolism in the brain (Barondes and Cohen, 1966; Flexner et al., 1967; Agranoff et al., 1967). The changes in protein metabolism in the brain presumably take place in the cell membrane. It is possible therefore that ACTH analogues act on membranes of cells in specific structures in the CNS, possibly in the same way as in isolated cells of the adrenal cortex (Seelig and Sayers, 1972), i.e., by inducing conformational changes that stimulate cyclic AMP production. This may result in stimulation of the metabolism of the cell, which is necessary to facilitate the formation of new synaptic connections.

REFERENCES

ADER, R., J. A. W. M. WEIJNEN, and P. MOLEMAN, 1972. Retention of a passive avoidance response as a function of the intensity and duration of electric shock. *Psychon. Sci.* 26: 125–128.

AGRANOFF, B. W., R. E. DAVIS, L. CASOLA, and R. LIM, 1967. Actinomycin D blocks formation of memory of shock-avoidance in goldfish. *Science* 158:1600–1601.

ANDERSON, D. C., W. WINN, and T. TAM, 1968. Adrenocorticotrophic hormone and acquisition of a passive avoidance response: A replication and extension. *J. Comp. Physiol. Psychol.* 66:497–499.

APPEL, S. H., W. DAVIS, and S. SCOTT, 1967. Brain polysomes: Response to environmental stimulation. *Science* 157:836–838.

APPLEZWEIG, M. H., and F. D. BAUDRY, 1955. The pituitary adrenocortical system in avoidance learning. *Psychol. Rep.* 1:417–420.

APPLEZWEIG, M. H., and G. MOELLER, 1959. The pituitary-adrenocortical system and anxiety in avoidance learning. *Acta Psychol.* 15:602–603.

BARONDES, S. H., and H. D. COHEN, 1966. Arousal and the conversion of "short-term" to "long-term" memory. *Proc. Nat. Acad. Sci. (USA)* 58:157–164.

BARRETT, R. J., N. J. LEITH, and O. S. RAY, 1971. The effects of pituitary-adrenal manipulations on time-dependent processes in avoidance learning. *Physiol. Behav.* 7:663–665.

BEATTY, P. A., W. W. BEATTY, R. E. BOWMAN, and J. O. GILCHRIST, 1970. The effects of ACTH, adrenalectomy and

dexamethasone on the acquisition of an avoidance response in rats. *Physiol. Behav.* 5:939–944.

BOHUS, B., and E. ENDRÖCZI, 1965. The influence of pituitary-adrenocortical function on the avoiding conditioned reflex activity in rats. *Acta Physiol. Acad. Sci. Hung.* 26:183–189.

BOHUS, B., and D. de WIED, 1966. Inhibitory and facilitatory effect of two related peptides on extinction of avoidance behavior. *Science* 153:318–320.

BOHUS, B., and D. de WIED, 1967a. Avoidance and escape behavior following medial thalamic lesions in rats. *J. Comp. Physiol. Psychol.* 64:26–29.

BOHUS, B., and D. de WIED, 1967b. Failure of α-MSH to delay extinction of conditioned avoidance behavior in rats with lesions in the parafascicular nuclei of the thalamus. *Physiol. Behav.* 2:221–223.

BOHUS, B., 1968. Pituitary ACTH release and avoidance behavior of rats with cortisol implants in mesencephalic reticular formation and median eminence. *Neuroendocrinology* 3: 355–365.

BOHUS, B., 1970. Central nervous structures and the effect of ACTH and corticosteroids on avoidance behavior: A study with intracerebral implantation of corticosteroids in the rat. In *Progress in Brain Research 32; Pituitary, Adrenal, and the Brain*, D. de Wied and J. A. W. M. Weijnen, eds. Amsterdam: Elsevier, pp. 171–184.

BOHUS, B., J. GRUBITS, G. KOVÁCS, and K. LISSÁK, 1970. Effect of corticosteroids on passive avoidance behavior of rats. *Acta Physiol. Acad. Sci. Hung.* 38:381–391.

BOHUS, B., 1971. Adrenocortical hormones and central nervous function: The site and mode of their behavioral action in the rat. In *Excerpta Med. Int. Congress Series* No. 219, V. H. T. James and L. Martini, eds. Amsterdam: Excerpta Medica, pp. 752–758.

BRAIN, P. F., 1971. Possible role of the pituitary-adrenocortical axis in aggressive behavior. *Nature (Lond.)* 233:489.

BRUSH, F. R., and S. LEVINE, 1966. Adrenocortical activity and avoidance learning as a function of time after fear conditioning. *Physiol. Behav.* 1:309–311.

CARDO, B., 1965. Rôle de certains noyaux thalamiques dans l'éboration et la conservation de divers conditionnements. *Psychol. France* 10:344–351.

CARDO, B., 1967. Effets de la stimulation du noyau parafasciculaire thalamique sur l'acquisition d'un conditionnement d'évitement chez le rat. *Physiol. Behav.* 2:245–248.

CELIS, M. E., S. TALEISNIK, and R. WALTER, 1971. Regulation of formation and proposed structure of the factor inhibiting the release of melanocyte-stimulating hormone. *Proc. Nat. Acad. Sci. (USA)* 68:1428–1433.

CHUNG, D., and C. H. LI, 1967. Adrenocorticotropins XXXVII. The synthesis of 8-lysine-ACTH 1-17NH$_2$ and its biological properties. *J. Amer. Chem. Soc.* 89:4208–4213.

CONNER, R. L., and S. LEVINE, 1969. The effects of adrenal hormones on the acquisition of signaled avoidance behavior. *Hormones Behav.* 1:73–83.

DEDMAN, M. L., T. H. FARMER, and C. J. O. R. MORRIS, 1955. Oxidation-reduction properties of adrenocorticotrophic hormone. *Biochem. J.* 59:xxii.

DELACOUR, J., D. A. FESSARD, and S. LIBOUBAN, 1966. Function of the two thalamic nuclei in instrumental conditioning in the rat. *Neuropsychologia* 4:101.

DELACOUR, J., 1970. Specific functions of a medial thalamic structure in avoidance conditioning in the rat. In *Progress in Brain Research 32; Pituitary, Adrenal, and the Brain*, D. de Wied and J. A. W. M. Weijnen, eds. Amsterdam: Elsevier, pp. 158–170.

DELLWEG, H., B. GERNER, and A. WACKER, 1968. Quantitative and qualitative changes in RNA of rat brain dependent on age and training experiment. *J. Neurochem.* 15:1109–1119.

de VELLIS, J., and D. INGLISH, 1968. Hormonal control of glycerol phosphate dehydrogenase in the rat brain. *J. Neurochem.* 15:1061–1071.

de WIED, D., 1964. Influence of anterior pituitary on avoidance learning and escape behavior. *Amer. J. Physiol.* 207:255–259.

de WIED, D., 1966. Inhibitory effect of ACTH and related peptides on extinction of conditioned avoidance behavior in rats. *Proc. Soc. Exp. Biol. Med.* 122:28–32.

de WIED, D., and B. BOHUS, 1966. Long term and short term effect on retention of a conditioned avoidance response in rats by treatment respectively with long acting pitressin or α-MSH. *Nature (Lond.)* 212:1484–1486.

de WIED, D., 1967. Opposite effects of ACTH and glucocorticosteroids on extinction of conditioned avoidance behavior. *Excerpta Medica Int. Congr. Series* No. 132:945–951.

de WIED, D., B. BOHUS, and H. M. GREVEN, 1968. Influence of pituitary and adrenocortical hormones on conditioned avoidance behavior in rats. In *Endocrinology and Human Behavior*, R. P. Michael, ed. New York: Oxford University Press, pp. 188–199.

de WIED, D., 1968. The anterior pituitary and conditioned avoidance behavior. *Excerpta Medica Int. Congr. Series* No. 184:310–316.

de WIED, D., and H. M. GREVEN, 1968. Opposite effect of structural analogues of ACTH on extinction of an avoidance response in rats by replacement of an L-amino acid or a D-isomer. Abstract 24, International Congress of Physiological Sciences, Washington, D.C., August 25–31, 1968. *Proc. Intern. Union Physiol. Sci.*, Vol. 7.

de WIED, D., 1969. Effects of peptide hormones on behavior. In *Frontiers in Neuroendocrinology*, W. F. Ganong and L. Martini, eds. New York: Oxford University Press, pp. 97–140.

de WIED, D., 1971. Pituitary-adrenal hormones and behavior. In *Normal and Abnormal Development of Brain and Behavior*, G. B. A. Stoelinga and J. J. van der Werff ten Bosch, eds. Boerhaave Series for Postgraduate Medical Education, Leiden, Netherlands: Leiden University Press, pp. 315–322.

de WIED, D., A. M. L. van DELFT, W. H. GISPEN, J. A. W. M. WEIJNEN, and TJ. B. van WIMERSMA GREIDANUS, 1972. The role of pituitary adrenal system hormones in active avoidance conditioning. In *Hormones and Behavior*, S. Levine, ed. New York: Academic Press, pp. 135–171.

FERRARI, W., G. GESSA, and L. VARGIU, 1963. Behavioral effects induced by intracisternally injected ACTH and MSH. *Ann. N.Y. Acad. Sci.* 104:330–343.

FLEXNER, B., J. B. FLEXNER, and R. B. ROBERTS, 1967. Memory in mice analyzed with antibiotics. *Science* 155:1377–1382.

GAITO, J., 1966. Macromolecules and brain function. In *Macromolecules and Behavior*. New York: Meredith.

GISPEN, W. H., and P. SCHOTMAN, 1970. Effect of hypophysectomy and conditioned avoidance behavior on macromolecule metabolism in the brain stem of the rat. In *Progress in Brain Research 32; Pituitary, Adrenal, and the Brain*, D. de Wied and J. A. W. M. Weijnen, eds. Amsterdam: Elsevier, pp. 221–235.

GISPEN, W. H., TJ. B. van WIMERSMA GREIDANUS, and D. de WIED, 1970a. Effects of hypophysectomy and ACTH$_{1-10}$ on responsiveness to electric shock in rats. *Physiol. Behav.* 5:143–147.

GISPEN, W. H., D. de WIED, P. SCHOTMAN, and H. S. JANSZ, 1970b. Effects of hypophysectomy on RNA metabolism in rat brain stem. *J. Neurochem.* 17:751–761.

GISPEN, W. H., P. SCHOTMAN, and E. R. de KLOET, 1972. Brain RNA and hypophysectomy: A topographical study. *Neuroendocrinology* 9:285–296.

GLASSMAN, E., 1967. *Molecular Approaches to Psychobiology*. New York: Dickenson.

GRAY, J. A., 1971. Effect of ACTH on extinction of rewarded behavior is blocked by previous administration of ACTH. *Nature (Lond.)* 229:52–54.

GRAY, J. A., A. R. MAYES, and M. WILSON, 1971. A barbiturate-like effect of adrenocorticotrophic hormone on the partial reinforcement acquisition and extinction effects. *Int. J. Neuropharmacol.* 10:223–230.

GREVEN, H. M., and D. de WIED, 1967. The active sequence in the ACTH molecule responsible for inhibition of the extinction of conditioned avoidance behavior in rats. *Europ. J. Pharmacol.* 2:14–16.

GREVEN, H. M., and D. de WIED, 1973. The influence of peptides derived from ACTH on performance structure activity studies. In *Progress in Brain Research*, E. Zimmerman, W. H. Gispen, B. H. Marks, and D. de Wied, eds. Amsterdam: Elsevier (in press).

GUPTA, S. L., and G. P. TALWAR, 1968. Effect of growth hormone on ribonucleic acid metabolism. The template activity of the chromatin and molecular species of ribonucleic acid synthetized after treatment with the hormone *Biochem. J.* 110:401–406.

GUTH, S., S. LEVINE, and J. P. SEWARD, 1971. Appetitive acquisition and extinction effects with exogenous ACTH. *Physiol. Behav.* 7:195–200.

HADŽOVIĆ, S., and D. de WIED, 1971. Central cholinergic pathways and the inhibitory effect of ACTH$_{1-10}$ on extinction of a pole jumping avoidance response. Abstr. First Congress Hungarian Society. Pharmacological Society, Budapest.

HENKIN, R. I., 1970. The effects of corticosteroids and ACTH on sensory systems. In *Progress in Brain Research 32; Pituitary, Adrenal and the Brain*, D. de Wied and J. A. W. M. Weijnen, eds. Amsterdam: Elsevier, pp. 270–294.

HODGES, J. R., and J. VERNIKOS-DANELLIS, 1962. Pituitary and blood corticotrophin changes in adrenalectomized rats maintained on physiological doses of corticosteroids *Acta Endocrin. (Kbh)* 39:79–86.

HOFMANN, K., R. ANDREATTA, H. BOHN, and L. MORODER, 1970. Studies on polypeptides XLV. Structure-function studies in the β-corticotropin series. *J. Med. Chem.* 13:339–345.

HYDÉN, H., and P. W. LANGE, 1965. A differentiation in RNA response in neurons early and late during learning. *Proc. Nat. Acad. Sci. (USA)* 53:946–952.

HYDÉN, H., and P. W. LANGE, 1968. Protein synthesis in the hippocampal pyramidal cells of rats during a behavioral test. *Science* 159:1370–1373.

HYDÉN, H., and P. W. LANGE, 1970. Brain cell protein synthesis specially related to learning. *Proc. Nat. Acad. Sci. (USA)* 65:898–904.

JAKOUBEK, B., B. SEMIGINOVSKY, M. KRAUS, R. ERDÖSSOVA, 1970. The alterations of protein metabolism of the brain cortex induced by anticipation stress and ACTH. *Life Sci.* 9:1169–1179.

KAMIN, L. J., 1963. Retention of an incompletely learned avoidance response: Some further analyses. *J. Comp. Physiol. Psychol.* 56:713–718.

KASPER-PANDI, PH., R. HANSING, and D. R. USHER, 1970. The effect of dexamethasone blockade of ACTH release on avoidance learning. *Physiol. Behav.* 5:361–363.

KAWAKAMI, M., T. KOSHINO, and Y. HATTORI, 1966. Changes in the EEG of the hypothalamus and limbic system after administration of ACTH, SU-4885 and ACh in rabbits with special reference to neurohumoral feedback regulation of pituitary-adrenal system. *Jap. J. Physiol.* 16:551–569.

KORÁNYI, L., C. BEYER, and C. GUZMÁN-FLORES, 1971a. Effects of ACTH and hydrocortisone on multiple unit activity in the forebrain and thalamus in response to reticular stimulation. *Physiol. Behav.* 7:331–335.

KORÁNYI, L., C. BEYER, and C. GUZMÁN-FLORES, 1971b. Multiple unit activity during habituation sleep-wakefulness cycle and the effect of ACTH and corticosteroid treatment. *Physiol. Behav.* 7:321–329.

KORNER, A., 1964 Regulation of the rate of synthesis of m-RNA by growth hormone. *Biochem. J.* 92:449–456.

LEONARD, B. E., 1969. The effect of sodium-barbitone, alone and together with ACTH and amphetamine, on the behavior of the rat in the multiple "T" maze. *Int. J. Neuropharmacol.* 8:427–435.

LEVINE, S., and L. E. JONES, 1965. Adrenocorticotropic hormone (ACTH) and passive avoidance learning. *J. Comp. Physiol. Psychol.* 59:357–360.

LEVINE, S., and F. R. BRUSH, 1967. Adrenocortical activity and avoidance learning as a function of time after avoidance training. *Physiol. Behav.* 2:385–388.

LISSÁK, K., E. ENDRÖCZI, and P. MEDGYESI, 1957. Somatisches Verhalten und Nebennierenrindentätigkeit. *Pflügers Arch.* 265:117–124.

LISSÁK, K., and B. BOHUS, 1972. Pituitary hormones and the avoidance behavior of the rat. *Int. J. Psychobiol.* 2:103–115.

MCEWEN, B. S., R. E. ZIGMOND, R. E. AZMITIA, JR., and J. M. WEISS, 1970. Steroid hormonal interaction with specific brain regions. In *Biochemistry of Brain and Behavior*, R. E. Bowman and S. P. Datta, eds. New York: Plenum Press, p. 123.

MILLER, R. E., and N. OGAWA, 1962. The effect of adrenocorticotrophic hormone (ACTH) on avoidance conditioning in the adrenalectomized rat. *J. Comp. Physiol. Psychol.* 55:211–213.

MOYER, K. E., 1958. The effect of adrenalectomy on anxiety-motivated behavior. *J. Genet. Psychol.* 92:11–16.

MURPHY, J. V., and R. E. MILLER, 1955. The effect of adrenocorticotrophic hormone (ACTH) on avoidance conditioning in the rat. *J. Comp. Physiol. Psychol.* 48:47–49.

OTSUKA, H., and K. INOUYE, 1964. Synthesis of peptides related to the N-terminal structure of corticotropin II. The synthesis of L-histidyl-L-phenylalanyl-L-arginyl-L-tryptophan, the smallest peptide exhibiting the melanocyte-stimulating and lipolytic activities. *Bull. Chem. Soc. Jap.* 37:1465–1471.

PAPPAS, B. A., and P. GRAY, 1971. Cue value of dexamethasone for fear-motivated behavior. *Physiol. Behav.* 6:127–130.

PASLEY, J. N., and J. J. CHRISTIAN, 1972. The effect of ACTH, group caging and adrenalectomy in *Peromyscus leucopus* with emphasis on suppression of reproductive function. *Proc. Soc. Exp. Biol. Med.* 139:921–925.

PFAFF, D. W., M. T. A. SILVA, and J. M. WEISS, 1971. Telemetered recording of hormone effects on hippocampal neurons. *Science* 172:394–395.

PHILLIPS, M. I., and N. DAFNY, 1971. Effect of cortisol on unit activity in freely moving rats. *Brain Res.* 25:651–655.

READING, H. W., and A. J. DEWAR, 1971. Effects of $ACTH_{4-10}$ on cerebral RNA and proteins metabolism in the rat. *Third Int. Meeting Int. Soc. Neurochem.*, Budapest, p. 199.

RICH, I., and R. THOMPSON, 1965. Role of the hippocamposeptal system, thalamus, and hypothalamus in avoidance conditioning *J. Comp. Physiol. Psychol.* 59:66–72.

SANDMAN, C. A., A. J. KASTIN, and A. V. SCHALLY, 1969. Melanocyte-stimulating hormone and learned appetitive behavior. *Experientia* 25:1001–1002.

SAWYER, CH. H., M. KAWAKAMI, B. MEYERSON, D. I. WHITMOYER, and J. LILLEY, 1968. Effects of ACTH, dexamethasone and asphyxia on electrical activity of the rat hypothalamus. *Brain Res.* 10:213–226.

SCHOTMAN, P., W. H. GISPEN, H. S. JANSZ, and D. de WIED, 1972. Effects of ACTH analogues on macromolecule metabolism in the brainstem of hypophysectomized rats. *Brain Res.* 46:349–362.

SEELIG, S., and G. SAYERS, 1972. $ACTH_{1-10}$ and $ACTH_{4-10}$ stimulate cyclic AMP production by isolated adrenal cells. *Fed. Proc.* 31:252.

SELYE, H., 1950. The physiology and pathology of exposure to stress. *Acta Inc. Montreal, Canada*, p. 6.

STEINER, F. A., 1970. Effects of ACTH and corticosteroids on single neurons in the hypothalamus. In *Progress in Brain Research 32; Pituitary, Adrenal, and the Brain*, D. de Wied and J. A. W. M. Weijnen, eds. Amsterdam: Elsevier, pp. 102–107.

TATA, J. R., and H. G. WILLIAMS-ASHMAN, 1967. Effects of growth hormone and tri-iodothyronine on amino acid incorporation by microsomal subfractions from rat. *Eur. J. Biochem.* 2:366–374.

THOMPSON, R., 1963. Thalamic structures critical for retention of an avoidance conditioned response in rats. *J. Comp. Physiol. Psychol.* 56:261–267.

TRACY, H. J., and R. A. GREGORY, 1964. Physiological properties of a series of synthetic peptides structurally related to gastrin I. *Nature (Lond.)* 204:935–938.

VAN DELFT, A. M. L., 1970. The relation between pretraining plasma corticosterone levels and the acquisition of an avoidance response in the rat. In *Progress in Brain Research 32; Pituitary, Adrenal and the Brain*, D. de Wied and J. A. W. M. Weijnen, eds. Amsterdam: Elsevier, pp. 192–199.

VANDERWOLF, C. H., 1964. Effect of combined medial thalamic and septal lesions on active-avoidance behavior. *J. Comp. Physiol. Psychol.* 58:31–37.

VAN WIMERSMA GREIDANUS, TJ. B., and D. de WIED, 1969. Effects of intracerebral implantation of corticosteroids on extinction of an avoidance response in rats. *Physiol. Behav.* 4:365–370.

VAN WIMERSMA GREIDANUS, TJ. B., 1970. The relation between pretraining plasma corticosterone levels and the acquisition of an avoidance response in the rat. In *Progress in Brain Research 32; Pituitary, Adrenal, and the Brain*, D. de Wied and J. A. W. M. Weijnen, eds. Amsterdam: Elsevier, pp. 185–191.

VAN WIMERSMA GREIDANUS, TJ. B., and D. de WIED, 1971. Effects of systemic and intracerebral administration of two opposite acting ACTH-related peptides on extinction of conditioned avoidance behavior. *Neuroendocrinology* 7:291–301.

Versteeg, D. H. G., W. H. Gispen, P. Schotman, A. Witter, and D. de Wied, 1972. Hypophysectomy and rat brain metabolism: Effect of synthetic ACTH analogs. In *Adv. Biochem. Psychopharmacol.* 6. New York: Raven Press, pp. 219–239.

Weijnen, J. A. W. M., and J. L. Slangen, 1970. Effects of ACTH-analogues on extinction of conditioned behavior. In *Progress in Brain Research 32; Pituitary, Adrenal and the Brain,* D. de Wied and J. A. W. M. Weijnen, eds. Amsterdam: Elsevier, pp. 221–233.

Weiss, J. M., B. S. McEwen, M. T. Silva, and M. Kalkut, 1970. Pituitary-adrenal alterations and fear responding. *Amer. J. Physiol.* 218:864–868.

Wertheim, G. A., R. L. Conner, and S. Levine, 1967. Adrenocortical influences on free-operant avoidance behavior. *J. Exp. Anal. Behav.* 10:555–563.

Woodbury, D. M., 1954. Effect of hormones on brain excitability and electrolytes. *Recent Progr. Hormone Res.* 10:65–104.

Zemp, J. W., J. E. Wilson, H. Schlesinger, W. O. Boggan, and E. Glassman, 1966. Brain function and macromolecules, I. Incorporation of uridine into RNA of mouse brain during short-term training experience. *Proc. Nat. Acad. Sci. (USA)* 55:1423–1431.

57 Macromolecules and Behavior: A Commentary

EDWARD GLASSMAN

ABSTRACT   This chapter is an attempt to point out the biochemical and behavioral difficulties that mar interpretation of the data in this field.

In addition some possible theoretical considerations are pointed out that may lead to fruitful research in the future.

RESEARCH attempting to relate macromolecules and learning has been extensively and adequately reviewed in recent years (Glassman, 1969; Jarvik, 1972; Horn, 1971; Ungar, 1972; Rose, 1970; Roberts and Matthysse, 1970). Suffice it to say that numerous chemical changes have been reported to be correlated with various training experiences. Almost all of these data are confounded because novel stimuli, such as flashing lights, activity, stress, shocks, etc., have been reported to cause similar chemical changes in the nervous system, and usually not enough behaviors have been examined to distinguish these psychological variables from the learning process. In addition, there are problems in the biochemical approaches that have been used. One of the purposes of this chapter is to discuss some of these problems in a critical and constructive way.

Problems of biochemical interpretations

Autoradiography is potentially one of the most elegant techniques for the study of chemical correlates of behavior because the response of large numbers of individual cells to an input stimulus can be monitored in a single preparation. To be effective, this technique requires the use of densitometry to avoid counting silver grains and computer analysis of the data to facilitate the comparison of the amount of radioactivity incorporated in cells or areas of various brains. Dr. Daniel Entingh has been working on such equipment for our laboratory (Entingh and Bernholz, unpublished).

Recently Rahmann (1973) has reported results that

EDWARD GLASSMAN   Department of Biochemistry, School of Medicine, University of North Carolina, Chapel Hill, North Carolina

constitute an extremely elegant example of autoradiographic analysis applied to the brain. Adult carp (*Carassius carassius*) were kept in darkness for 10 days, after which they were injected intraperitoneally with [^{3}H]-histidine, curarized, and one eye only was exposed for 75 min to 15 sec of light alternating with 45 sec of darkness. The fish were immediately sacrificed and autoradiograms were prepared of histological sections through the optic tecti. The density of grains over the stimulated contralateral tectum was significantly greater than that over the non-stimulated ipsilateral tectum. Of even greater interest, exposure of one eye to a vertical stripe of light 2 mm wide produced an increased density of silver grains over a narrow stripe in the stimulated tectum, while two vertical stripes of light produced two such stripes. Similar results have also been reported by Rahmann's associates in the guppy (Skrzipek, 1969) and the frog (Wegener, 1970). This elegant experiment indicates the exciting kind of data that can be generated in complex nervous systems using this approach.

It should be noted, however, that the biochemical conclusions to be drawn from autoradiographic techniques using radioactive tracers are not unequivocal. For example, the timing of most experiments is long enough for much of the radioactive amino acid to be metabolized to other substances, and therefore it is not clear whether the detected radioactivity is in protein or in some other compound. In addition, the increase in radioactivity could be due to increased synthesis, decreased destruction (see chapters by Schimke and by Goldberg in this volume), or the availability of radioactive precursor. This problem has been previously discussed (Glassman, 1969), but its importance is increasingly evident (see especially Baskin et al., 1972).

The problem of monitoring the availability of the radioactive precursor is particularly acute, because there seems little chance of measuring it simply in an intact animal. Changes in the amounts of radioactive precursor in individual cells or areas of the nervous system can occur because of localized changes in blood flow, changes in cellular permeability to the radioactive compound, changes in endogenous rates of synthesis in individual cells, and localized differences due to unavoidable variations in the injection. The problem of localized changes in blood flow is of particular concern in view of the report by Sokoloff (1961) that visual stimulation can cause an increase of blood flow in the visual cortex of light-deprived rats.

One can partially correct for the availability of a radioactive precursor by measuring the radioactivity present in the precursor or in other appropriate small molecules of the cells when the animals are sacrificed, and thereby derive a pool correction factor. This will monitor some variations in injection procedure, and possibly some changes in blood flow, cellular permeability, and changes in endogenous synthesis. Any pool correction factor should be viewed with caution, however, unless it is measured throughout the entire time of the incorporation period, and it can be shown that the behavioral stimulation has no effect on the radioactivity in the pool. Even if these criteria are met, the problem of compartmentation of metabolic pools within a cell, particularly with respect to the site of synthesis, makes gross pool correction factors inadequate to monitor the immediate precursor pool for any macromolecule.

The inability to derive data on this crucial point is probably the reason that the problem of pool correction factors has not been the subject of more extensive investigation. Most investigators have contented themselves with a crude pool estimate, or none at all. It is of interest that it is extremely difficult to obtain a pool correction factor in autoradiographic analysis (but see Watson, 1965, and Quevedo et al., 1971, for attempts to solve this problem). Thus one cannot conclude that results obtained using autoradiography are due only to altered macromolecular synthesis. This is not meant to detract, however, from the importance and interest of finding changes in incorporation limited to specific cells and regions of the nervous system but rather to point out where the biochemical conclusions are equivocal and where further experimental clarification is necessary. Because of the importance of evaluating the validity of a pool correction factor, it is imperative that it is clearly presented in papers that present data using radioactive tracers.

These ideas can be applied to many investigations using radioactive tracers. For example, Zemp et al. (1966) reported increased incorporation of uridine into brain RNA due to avoidance training using the radioactivity in UMP as a pool correction factor; but Entingh et al. (in preparation) have shown that a possible alternative interpretation is that the radioactivity in UMP is decreased as a result of the experience. Another example is the work of Hydén and Lange (1968, 1970) who reported increased incorporation of [^{3}H]-leucine into specific proteins of CA3 cells of the hippocampus of rats undergoing a change in handedness, but who derived a soluble pool correction factor from a different group of cells (see also Comment 1 of Bowman and Harding, 1969, who discuss a similar point). In addition, the time of sacrifice following the injection in these experiments is so long that it seems likely that most of the radioactivity is no longer in leucine but in other substances. Since increased incorporation into specific proteins is involved, one possible correction factor might be the amount incorporated into other proteins (see Gisiger and Gaide-Huguenin (1969) and Kerkut et al. (1970) for RNA; and Emson et al. (1971), Wilson (1971), Gainer (1972a, b, c), and Strumwasser (this volume) for protein). Hydén (personal communication) has stated that the techniques used in his laboratory had been changed so some of these comments may not apply to some of his more recent work (see Hydén and Lange, 1972a, 1972b).

Other investigators have indicated their concern with this problem. Rose (1972) reported on the effect of visual experience on amino acid pools of rat brain. Baskin et al. (1972) reported that the incorporation of radioactivity from orotic acid into cytidylic acid is decreased in goldfish brain by the CO_2 produced by the activity of the fish, with a resulting decrease of radioactive cytidine in RNA. Electrical stimulation of isolated nerve tissue has been reported to produce changes in amino acid pools (Jones and McIlwain, 1971; Jones, 1972; Jones and Banks, 1970a, 1970b; McBride and Klingman, 1972; Orrego and Lipmann, 1967) and in RNA pools (Orrego, 1967; Prives and Quastel, 1969). Wilson and Berry (1972) report that the amount of incorporation of uridine into RNA of R2 cells of *Aplysia* is dependent on the external uridine concentration, an observation in agreement with the data of Itoh and Quastel (1969) using rat cortex slices.

One way to avoid the problems associated with valid pool correction factors is not to use radioactive tracers but to measure changes in the amount of the substance directly. For example, Haljamäe and Lange (1972) have recently reported that rats undergoing a change in handedness show increases in the amount of S-100 protein in the CA3 cells of the hippocampus. Furthermore this increased amount of S-100 appears to migrate differently during microdisc gel electrophoresis, suggesting a changed conformation. These workers also report a significant increase of calcium in these cells but no change in sodium or potassium. Since calcium has been reported to cause conformational changes in S-100 (Calissano et al., 1969), these workers suggest that the increased calcium in these cells combines with S-100, and a conformational form with increased stability results. Haywood and Rose (1970) reported an increase in RNA polymerase following exposure of unimprinted, sensitive chicks to a flashing light. The increase occurred in the same part of the brain that increased incorporation of radioactive precursor into RNA, and protein have been reported under similar behavioral stimulation (Bateson et al., 1972). Kerkut et al. (1970) and Emson et al. (1971) reported changes in cholinesterases with learning in invertebrate ganglia (but see Woodson et al., 1972). Machlus et al (1973a, 1973b) have reported increased radioactive phosphate in nuclear proteins of rat brain as a result of a short training experience. This effect of experience was also correlated with an increase in the molar ratios of phosphoserine compared to serine in hydrolysates of these proteins; this direct measurement avoided the problem of an adequate pool correction factor since no radioactivity was involved. Finally, there are reports of the effects of experience on polysomes that also do not involve radioactive tracers (Appel et al., 1967; MacInnes et al., 1970; Uphouse et al., 1972a, 1972b). These comments are not meant to downgrade the use of radioactive tracers, because the use of such techniques provides excellent clues to possible important chemical changes in small areas of nerve tissue and are important in any research strategy. Nonetheless their exclusive use does not permit an unequivocal conclusion that only synthesis is being monitored.

Problems of behavioral interpretations

The causative stimuli in this research are difficult to ascertain. It may be that the learning per se or the special stresses, emotions, and motivations that accompany learning are related to the observed chemical responses to training experiences, but it is also possible that the changes are due to the other stimuli that accompany the training experience. Sensory stimulation and stressful situations have been reported to produce changes in macromolecules in the nervous system, and unless adequate behavioral experiments are carried out, it is difficult to eliminate such psychological variables as the cause of chemical changes in the brain.

A major difficulty in quantitative correlation is the problem of assaying the amount of memory stored after a training experience. Such assays depend on the performance of an animal and thus are sensitive to such diverse factors as the motivation and attention of the animal, its state of health and activity, and a number of other emotional and physical factors. Thus an animal may not perform well if it is not motivated, or too sick, tired, or frightened, or if it has a memory deficit. Even if extensive experiments are carried out, it is very difficult to establish why an animal is not performing well.

Another problem arises from exaggerated concerns as to whether such changes are directly associated with

learning and memory processes, or whether they are due to *nonspecific* responses. The difficulty, mostly semantic, is that nonspecific responses do not exist; they are merely responses whose cause is unknown. It would probably be more productive to define such responses in definite terms, and to attempt to classify them as encoding-specific, training-specific, consolidation-specific, imprinting-specific, stimulus-specific, emotion (arousal)-specific, stress-specific, attention-specific, performance-specific, etc. For example, Rahmann's autoradiographic results described above are probably due to a stimulus-specific response in the reacting cells. The data of Bateson et al. (1972) may be due to an imprinting-specific response, but the behavioral data do not rule out a stimulus-specific response to flickering light. In the change-of-handedness experiments, untrained rats used for comparison with the trained animals are always rats using their preferred paws (Hydén and Lange, 1968, 1970a, b). Thus these data are inadequate to distinguish between training-specific responses, or those due to the other behavioral- and psychological-specific responses listed above. This is extremely important in view of the finding by Rees and Brogan (see Dunn et al., this volume) that an increased incorporation into rat brain protein can be observed 20 to 30 min following a wide variety of novel stimuli that do not involve training. The data of Adair et al. (1968a, 1968b) show training-specific increased incorporation of radioactive uridine into brain polysomes, but the behavioral data are inadequate to determine whether this is involved with encoding or with the specific motivational or emotional changes associated with the training experience. This is equally true for all other data involving training experiences. Machlus et al. (1973a, 1973b) demonstrate a more complex behavioral interaction in that the increase of radioactive phosphate in brain nuclear protein is not only a training-specific response, but it also occurs after a training-specific reminder, an effect that Adair et al. (1968a) did not find.

These kinds of data indicate that behavioral analysis to determine the psychological variables involved is extremely crucial to understanding these phenomena. No one behavior is an adequate control; indeed, the term "control" should be avoided. The effects of many related experiences must be studied before the specificity of a chemical response can be evaluated.

Studies on memory consolidation

In order to study macromolecular correlates of memory processes in vertebrates, it would be helpful to know which brain area is involved, when the chemical change occurs, and which chemicals to study. At one time it was felt that the data on memory consolidation would provide important clues for the neurochemist. It is no longer clear that this is the case.

Various treatments immediately after training to *localized* areas of the brain can cause deficits or facilitation of performance when the animals are tested at a later time (for example, see Grossman, 1969; Grossman and Mountford, 1964; Stein and Chorover, 1968; Wyers et al., 1968; Mahut, 1962; Thompson, 1958; Kesner and Doty, 1968; Kesner and Conner, 1972; Glickman, 1958; Zornetzer, 1972; Zornetzer and McGaugh, 1972; Denti et al., 1970; Daniels, 1971a; Flexner, 1967; Vardaris and Schwartz, 1971; Lidsky and Slotnick, 1970; Herz and Peeke, 1971; Peeke and Herz, 1971; Wyers and Deadwyler, 1971; Hudspeth and Wilsoncroft, 1969; Paolino and Levy, 1971; Gold et al., 1971; Dorfman and Jarvik, 1968; Erickson and Patel, 1969). Much of this is excellent data and implicates many subcortical structures. The pattern that emerges, however, is far too diffuse to provide clear clues to the neurochemist as to precisely where to look for chemical changes that might be associated with the formation of long-term memory.

Clues as to which chemicals might be involved in the formation of long-term memory are not easily found when the chemical effects of amnestic agents are analyzed. Almost all agents used so far have a wide variety of effects that confound attempts to postulate a common cause. This is true even if one makes allowances for a multistaged complex process. One possible common effect of most amnestic agents may be to reduce the level of arousal long postulated to be a requirement for memory consolidation, although another possibility may be their ability to inhibit RNA or protein synthesis (see Glassman, 1969).

The time at which the neurochemist should look for chemical correlates of the formation of long-term memory seems clear, because most reports indicate the process starts during or soon after training. The exact time is under dispute, and even the nature of the process in terms of defective storage or defective retrieval mechanisms is not agreed on (see Quartermain and McEwen, 1970; Squire and Barondes, 1972). It has been reported that long-term memory formation following an aversive one-trial pecking task in chicks is decreased by ouabain, lithium, or copper ions, but only if they are injected just prior to the training trial, whereas cycloheximide is effective if injected just before or just after the task (Mark and Watts, 1971; Watts and Mark, 1971). Since one common effect of ouabain, lithium, and copper is to inhibit the sodium pump in nerve membranes, it was concluded that this process is essential for the formation of long-term memory before there is a requirement for protein synthesis. The fact that such compounds can affect the formation of long-term memory should stimulate further research into the whole range of possible chemical amnestic agents.

Whether an expanded list will clarify or add to the confusion remains to be seen.

Some theoretical considerations

There are a number of possible roles of macromolecules in brain function. One possibility is that the macromolecules reported as biochemical correlates to training experiences are involved in the restoration of the cellular components used during nerve activity. This nerve activity could be directly associated with the learning process per se but might be involved with motivational, emotional, or other phenomena that accompany behavioral stimulation.

It does not seem probable that these macromolecules are themselves encoding the experiential information; but another possibility is that such chemical changes are related to those processes that affect the response of neurons to impulses from neighboring neurons by changing the connectivity between neurons so that new pathways become functional. This could occur by changes in the effectiveness of the neurotransmitter, e.g., by increasing the amount released, by decreasing the activity of enzymes that inactivate it, or by increasing the number or effectiveness of the receptor sites on the postsynaptic neuron, or by other means. Thus, the observed changes in macromolecules may reflect the important role of regulating interneuronal connectivity. It is tempting to speculate that they are part of the early steps in the storage of long-term memory, a process that may be dependent on RNA and protein synthesis (Flexner et al., 1967; Lande et al., 1972; Agranoff, 1972; Daniels, 1971b; Squire and Barondes, 1972).

Figure 1 summarizes these ideas. A training experience has many effects on an animal. Increased electrophysiological activity (information flow) has been reported (see John, 1967), and it is thought that through associative processes of various types involving feedbacks and special

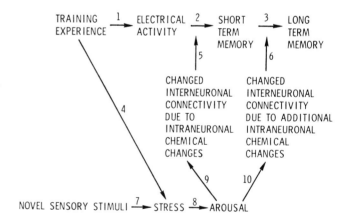

FIGURE 1   Some effects of a training experience.

instructions to a variety of cells, intracellular chemical changes take place that produce connectivity changes between neurons. The final result is the formation of new pathways that underlie short-term memory, and later, long-term memory. Thus the interneuronal events that encode memory (steps 2 and 3) are distinct from supportive intraneuronal events involving chemical changes (steps 5 and 6). It has long been felt that the formation of long-term memory probably depends on a type of arousal that stimulates a sequence of chemical events that are independent of short-term memory (see, for example, Barondes and Cohen, 1968). The arousal may also be produced by other experiences (step 7). This would account for many of the chemical changes caused by novel sensory stimulation. These molecules may be related to the peptide(s) postulated to influence the maintenance and retrieval of memory by de Wied (this volume).

The basic questions that need to be answered in order to evaluate these ideas include the way electrical activity is transduced into the chemical changes necessary to regulate or change interneuronal connectivity, the sequence of the biochemical changes involved, and the primary cellular trigger for these events. Research involving isolated ganglia, cortical slices, giant neurons of invertebrates and other isolated nervous system components can provide clues to the answers of some of these questions. One striking finding is that many chemical changes within a neuron are correlated with the stimulation of the neuron by presynaptic stimulation and do not seem to occur when the axon of the neuron is stimulated antidromically (Larrabee and Leicht, 1965; Hokin et al., 1960; Stoller and Wayner, 1968; McAffee et al., 1971). This effect of presynaptic stimulation has been reported in relation to the increase of radioactive phosphate in phosphatidylinositol of mammalian sympathetic ganglia (Larrabee and Leicht, 1965; Larrabee, 1968; Hokin et al., 1960; Hokin, 1965, 1966); the increased incorporation of uridine into RNA of *Aplysia* cells (Kernell and Peterson, 1970; but see Wilson and Berry, 1972) and rat superior cervical sympathetic ganglion (Gisiger and Gaide-Huguenin, 1969; Gisiger, 1971); the increase of cyclic AMP of rabbit superior cervical sympathetic ganglion (McAffee et al., 1971) or *Aplysia* abdominal ganglion (Cedar et al., 1972a); and the changes of activity of various enzymes involved in brain amine metabolism (Axelrod, this volume). That the action potential generated by antidromic stimulation and stimulation across a synapse has different effects is also suggested by the report that antidromic stimulation of an inhibitory neuron does not lead to the expected inhibition of other neurons (Mulloney and Selverston, 1972).

Many chemical changes induced by presynaptic stimulation can be duplicated by applying neurotrans-mitter (Cedar and Schwartz, 1972b; Prives and Quastel, 1969; Hokin, 1969b; Hokin et al., 1960; Larrabee and Leicht, 1965; Shimizu et al., 1970; Kakiuchi and Rall, 1968a, 1968b; Klainer et al., 1962; Gilman and Nirenberg, 1971; McAffee et al., 1971; Kuo et al., 1972). Furthermore, these chemical changes are blocked by agents that prevent the transmitter from being released (Cedar and Schwartz, 1972a) or from attaching to the receptor site (Prives and Quastel, 1969; Gisiger and Gaide-Huguenin, 1969; Gisiger, 1971; Larrabee and Leicht, 1965) but not necessarily by treatments that affect the action potential (Cedar and Schwartz, 1972a; Gisiger, 1971).

It would therefore appear that the initial trigger for a variety of chemical changes in a neuron depends on the attachment of the neurotransmitter from the presynaptic cell, as shown in Figure 2. In this figure, sequence IV shows the changes involved in the generation of the action potential of the cell. It is well known that this involves temporary changes in membrane permeability to ions, but as indicated in Figure 2, sequence III, there are also changes in neuronal permeability to other ions, amino acids, nucleosides, and other compounds. This emphasizes again the importance of determining pool correction factors in the same cells in which radioactivity in macromolecules is being determined.

Reaction sequence V of Figure 2 describes chemical events that are well known in many types of cells, and that may underlie the molecular control of neuronal connectivity. The activation of protein kinases by cyclic-AMP or cyclic-GMP (step b) eventually leads to conformational changes in proteins (step d) via their phosphorylation (step c). If the conformational changes take place in synaptic proteins (step e^1), then one can postulate rapid, direct effects on synaptic properties resulting in changes in neuronal connectivity with eventual memory formation. An alternative idea is that conformational changes take place in nuclear proteins (step e), a process that leads to gene activation, RNA synthesis, and protein synthesis (steps f and g). This protein can be involved in processes involved with replenishing the chemicals of the neuron (step h^1) or, alternatively, may have special neuronal functions (step h) at the synapse (step j), where it regulates synaptic properties such as connectivity, etc. (steps k, l, m). It is possible that during training, conformational changes in synaptic proteins occur rapidly to temporarily change connectivity associated with the formation of short-term memory, but that the events following conformational changes in the nuclear proteins are necessary for the permanent connectivity changes that underlie the formation of long-term memory.

There are many lines of research that suggest these ideas have merit, but none are sufficient to prove them

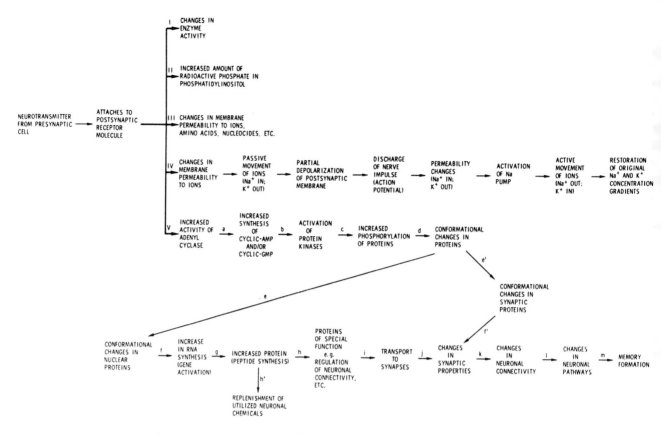

FIGURE 2   Some chemical events that follow the attachment of a neurotransmitter to a neuron.

true, or that any known chemical events do indeed regulate interneuronal connectivity, pathway formation, or memory formation. The reason that attention has focused on conformational changes is that short-term nervous system function is not affected by extensive inhibition of protein synthesis. For example, the formation of short-term memory is not impaired by doses of cycloheximide, puromycin, or actinomycin that inhibit the formation of long-term memory (see Glassman, 1969). In addition, a number of researchers using invertebrate preparations report that sustained neurophysiological activity can take place in the virtual absence of RNA or protein synthesis (Schwartz et al., 1971; Edstrom and Grampp, 1965; Bondeson et al., 1967; Kandel, this volume). It has been reported (see Schwartz et al., 1971; Castellucci et al., 1972) that the generation of the action potential, the resting membrane potential, posttetanic potentiation, inhibitory and excitatory synaptic potentials, and short-term habituation can take place in the virtual absence of protein synthesis in *Aplysia*. Whether long term phenomena in this organism are dependent on protein synthesis remains to be determined.

The way in which conformational changes in molecules can regulate interneuronal connectivity is not known. One can postulate that a chemical located at the synapse or elsewhere in the pre- or postsynaptic neuron is critical for connectivity, and that the control of connectivity is exerted by the conformational state of this chemical. As shown in Figure 3, conformation state A is stable, but connectivity is not facilitated. If during training a pathway or network is activated in which this synapse is important, the molecule is shifted into conformational state B. Conformational state B is unstable, but it changes the properties of the neuron so that a temporary change in connectivity occurs. The new pathway or network can be considered a part of the mechanism underlying short-term memory. The molecule in conformation state B can revert back to conformation A, in which case connectivity at this synapse is again as it was, and the memory is lost or is no longer easily retrievable. Alternatively, if certain factors are stimulated in the animal undergoing training, the molecule in conformation B is converted to a new stable form, conformation C, so that connectivity is permanently changed. Long-term memory is now possible. Since only molecules in conformation B can be converted to conformation C, only those pathways or networks that have been previously selected as providing adaptive behavior will be affected and made permanent. Alternatively, new molecules can be synthesized to make the connectivity changes permanent. Figure 3 is based on the idea that the pathways or networks underlying short- and long-term memory are the same, or at least overlap to a great extent (see Kandel, this volume). Much work is necessary to determine the validity of ideas like this, but it has considerable heuristic value. Thinking along such lines has led to the study of the effects of behavior on the metabolism of membrane constituents, particularly those found in synapses (see Dunn et al., this volume).

It should be pointed out that phosphorylation is not the only way to achieve conformational changes in proteins. Others include the enzymatically catalyzed formation of covalent bonds with methyl groups, acetyl groups, and carbohydrates. The presence of substrates, coenzymes, ions, and a variety of hormones and regulators will also cause conformational changes, as will selective removal of amino acids or small peptides. Barondes (1970) has reviewed the possible role of brain glycoproteins in interneuronal recognition.

Conclusions

One purpose of this article is to attempt to evaluate the significance of the reported effects of behavior and training experiences on macromolecules in the nervous system. A detailed discussion of the role of these macromolecules in the learning processes is certainly premature; there is no clear evidence that these macromolecules play a direct role in encoding, and cause and effect relationships in this area are still difficult to prove. The current problems that seem important are the aspects of the environment or the behavior that can trigger chemical responses in the nervous system, the physiological and biochemical processes that convey the information from outside the animal to the nerve cells that are responding, and the significance of the chemical response in terms of the functional role that these macromolecules play in the cells of the nervous system. Even if no aspect of the learning process is involved, the fact that behavior can affect macromolecules in these ways should be of great interest to those studying nervous system function. The role these macromolecules play in the nervous system may be only to maintain cellular health and structure and to enable the cellular machinery to function. If so, it is important to know this fact so that the solutions to the problem of the unique mechanisms of the nervous system in regulating behavior can be sought elsewhere.

Certain specific problems can be emphasized. The first concerns the significance of the increased amounts of radioactive phosphate in phosphatidylinositol with presynaptic stimulation and its possible role in reaction V of Figure 2. Hokin (1969b) has suggested this phenomenon is related to memory processes (see also review by Hokin, 1969a, and Durell et al., 1969). One approach would be to determine whether behavioral experiences, particularly training, will affect this reaction.

A second question concerns the report of a training-specific increase in the amount of radioactive phosphate in nuclear proteins (Machlus et al., 1973a, 1973b) and in

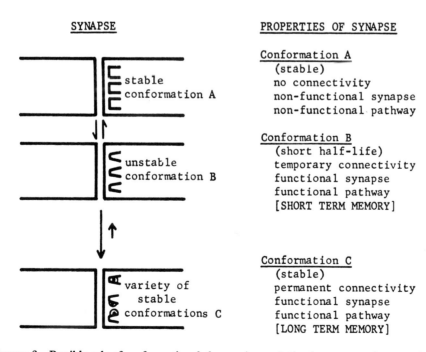

SYNAPSE

PROPERTIES OF SYNAPSE

stable conformation A

unstable conformation B

variety of stable conformations C

Conformation A
(stable)
no connectivity
non-functional synapse
non-functional pathway

Conformation B
(short half-life)
temporary connectivity
functional synapse
functional pathway
[SHORT TERM MEMORY]

Conformation C
(stable)
permanent connectivity
functional synapse
functional pathway
[LONG TERM MEMORY]

FIGURE 3   Possible role of conformational changes in regulating interneuronal connectivity.

synaptosomal proteins (Glassman et al., 1973) and their possible relation to sequence V of Figure 2. Jones and Rodnight (1971) have reported a similar phenomenon when cortical slices of guinea pig brain were electrically stimulated, and it would be of interest to determine whether such changes are dependent on presynaptic or antidromic stimulation.

Another question concerns the effects of nerve activity in single neurons on the incorporation of various precursors into macromolecules while the precursor pool and changes in cellular permeability are carefully monitored. Possibly the application of techniques for directly monitoring biochemical changes during nerve function will be of use here (see, for example, potential applications described by Aubert et al., 1964; Maitra et al., 1964; Tasaki et al., 1969, 1972; Cummins and Bull, 1971; Clark and Strickholm, 1971; Moore and Wetlaufer, 1971; Baum and Ward, 1971; Kohen and Kohen, 1966; Cummins, 1971; Llinas et al., 1972). The combination of such techniques with inhibitors that specifically block the various steps involved in the generation of the action potential (sequence III, Figure 2) might be very revealing.

Finally the role of arousal in these processes needs elucidation. The approach of Aprison et al. (1968) seems particularly suited to studies on the effect of experience on brain amines (see also Serota, 1971, Serota et al., 1972; Randt et al., 1971).

The problem of molecular control of long term changes in interneuronal connectivity is a challenge; its solution may provide insights into nervous system function that can be obtained in no other way.

ACKNOWLEDGMENTS   I am deeply indebted to the members of the Division of Chemical Neurobiology for many fruitful discussions over the years. I am particularly grateful to Dr. Daniel Entingh, Dr. Adrian Dunn and Dr. John Wilson who critically read earlier versions of this manuscript.

REFERENCES

ADAIR, L. B., J. E. WILSON, and E. GLASSMAN, 1968a. Brain function and macromolecules. IV. Uridine incorporation into polysomes of mouse brain during different behavioral experiences. *Proc. Nat. Acad. Sci. USA* 61:917–922.

ADAIR, L. B., J. E. WILSON, J. W. ZEMP, and E. GLASSMAN, 1968b. Brain function and macromolecules. III. Uridine incorporation into polysomes of mouse brain during short-term avoidance conditioning. *Proc. Nat. Acad. Sci. USA* 61:606–613.

AGRANOFF, B. W., 1972. Further studies on memory formation in the goldfish. In *The Chemistry of Mood, Motivation, and Memory*, J. L. McGaugh, ed. New York: Plenum Press, pp. 175–185.

APPEL, S. H., W. DAVIS, and S. SCOTT, 1967. Brain polysomes: Response to environmental stimulation. *Science* 157:836–838.

APRISON, M. H., T. KARIYA, J. N. HINGTGEN, and M. TORU, 1968. Neurochemical correlates of behavior: Changes in acetylcholine, norepinephrine and 5-hydroxytryptamine concentrations in several discrete brain areas of the rat during behavioural excitation. *J. Neurochem.* 15:1131–1139.

AUBERT, X., B. CHANCE, and R. D. KEYNES, 1964. Optical studies of biochemical events in the electric organ of *Electrophorus*. *Proc. R. Soc. Lond. (Biol.)* 160:211–245.

BARONDES, S. H., 1970. Multiple steps in the biology of memory. In *The Neurosciences: Second Study Program*, F. O. Schmitt, ed. New York: The Rockefeller University Press, pp. 272–278.

BARONDES, S. H., and H. D. COHEN, 1968. Arousal and the conversion of "short-term" to "long term" memory. *Proc. Nat. Acad. Sci. USA* 61:923–929.

BASKIN, F., F. R. MASIARZ, and B. W. AGRANOFF, 1972. Effect of various stresses on the incorporation of [³H]orotic acid into goldfish brain RNA. *Brain Res.* 39:151–162.

BATESON, P. P. G., G. HORN, and S. P. R. ROSE, 1972. Effects of early experience on regional incorporation of precursors into RNA and protein in the chick brain. *Brain Res.* 39:449–465.

BAUM, G., and F. B. WARD, 1971. General enzyme studies with a substrate-selective electrode: Characterization of cholinesterases. *Anal. Biochem.* 42:487–493.

BONDESON, C., A. EDSTROM, and A. BEVIZ, 1967. Effects of different inhibitors of protein synthesis on electrical activity in the spinal cord of fish. *J. Neurochem.* 14:1032–1034.

BOWMAN, R. E., and G. HARDING, 1969. Protein synthesis during learning. *Science* 164:199–201.

CALISSANO, P., B. W. MOORE, and A. FRIESEN, 1969. Effect of calcium ion on S-100, a protein of the nervous system. *Biochem.* 8:4318–4326.

CASTELLUCCI, V. F., E. R. KANDEL, and J. H. SCHWARTZ, 1972. Macromolecular synthesis and the functioning of neurons and synapses. In *Structure and Function of Synapses*, G. D. Pappas and D. P. Purpura, eds. New York: Raven Press, pp. 193–219.

CEDAR, H., E. R. KANDEL, and J. H. SCHWARTZ, 1972a. Cyclic AMP in the nervous system of *Aplysia californica*. I. Increased synthesis in response to synaptic stimulation. *J. Gen. Physiol.* 60:558–569.

CEDAR, H., and J. H. SCHWARTZ, 1972b. Cyclic AMP in the nervous system of *Aplysia californica*. II. Effect of serotonin and dopamine. *J. Gen. Physiol.* 60:570–587.

CLARK, H. R., and A. STRICKHOLM, 1971. Evidence for a conformational change in nerve membrane with depolarization. *Nature (Lond.)* 234:470–471.

CUMMINS, J. T., 1971. Spectral changes in respiratory intermediates of brain cortex in response to depolarizing pulses. *Biochim. Biophys. Acta* 253:39–45.

CUMMINS, J. T., and R. BULL, 1971. Spectrophotometric measurements of metabolic responses in isolated rat brain cortex. *Biochim. Biophys. Acta* 253:29–38.

DANIELS, D., 1971a. Acquisition, storage, and recall of memory for brightness discrimination by rats following intracerebral infusion of acetoxycycloheximide. *J. Comp. Physiol. Psychol.* 76:110–118.

DANIELS, D., 1971b. Effects of actinomycin D on memory and brain RNA synthesis in an appetitive learning task. *Nature (Lond.)* 231:395–397.

DENTI, A., J. L. MCGAUGH, P. W. LANDFIELD, and P. G. SHINKMAN, 1970. Effects of posttrial electrical stimulation of the mesencephalic reticular formation on avoidance learning in rats. *Physiol. Behav.* 5:659–662.

DORFMAN, L. J., and M. E. JARVIK, 1968. Comparative amnesic

effects of transcorneal and transpinnate ECS in mice. *Physiol. Behav.* 3:815–818.

DURELL, J., J. T. GARLAND, and R. O. FRIEDEL, 1969. Acetylcholine action: Biochemical aspects. *Science* 165:862–866.

EDSTROM, J.-E., and W. GRAMPP, 1965. Nervous activity and metabolism of ribonucleic acids in the crustacean stretch receptor neuron. *J. Neurochem.* 12:735–741.

EMSON, P., R. J. WALKER, and G. A. KERKUT, 1971. Chemical changes in a molluscan ganglion associated with learning. *Comp. Biochem. Physiol.* 40B:223–239.

ERICKSON, C. K., and J. B. PATEL, 1969. Facilitation of avoidance learning by posttrial hippocampal electrical stimulation. *J. Comp. Physiol. Psychol.* 68:400–406.

FLEXNER, L. B., 1967. Memory in mice dissected with antibiotics. *Amer. J. Dis. Child.* 114:574–580.

FLEXNER, L. B., J. B. FLEXNER, and R. B. ROBERTS, 1967. Memory in mice analyzed with antibiotics. *Science* 155:1377–1383.

GAINER, H., 1972a. Effects of experimentally induced diapause on the electrophysiology and protein synthesis patterns of identified molluscan neurons. *Brain Res.* 39:387–402.

GAINER, H., 1972b. Electrophysiological behavior of an endogenously active neurosecretory cell. *Brain Res.* 39:403–418.

GAINER, H., 1972c. Patterns of protein synthesis in individual, identified molluscan neurons. *Brain Res.* 39:369–385.

GILMAN, A. G., and M. NIRENBERG, 1971. Regulation of adenosine 3′,5′-cyclic monophosphate metabolism in cultured neuroblastoma cells. *Nature (Lond.)* 234:356–357.

GISIGER, V., 1971. Triggering of RNA synthesis by acetylcholine stimulation of the postsynaptic membrane in a mammalian sympathetic ganglion. *Brain Res.* 33:139–146.

GISIGER, V., and A.-C. GAIDE-HUGUENIN, 1969. Effect of preganglionic stimulation upon RNA synthesis in the isolated sympathetic ganglion of the rat. *Prog. Brain Res.* 31:125–129.

GLASSMAN, E., 1969. The biochemistry of learning: An evaluation of the role of protein and nucleic acids. *Ann. Rev. Biochem.* 38:605–646.

GLASSMAN, E., W. H. GISPEN, R. PERUMAL, B. MACHLUS, and J. E. WILSON, 1973. The effect of short experiences on the incorporation of radioactive phosphate into synaptosomal and non-histone acid-extractable nuclear proteins from rat and mouse brain., *5th Int. Congr. Pharmacol.* 4: (in press).

GLICKMAN, S. E., 1958. Deficits in avoidance learning produced by stimulation of the ascending reticular formation. *Can. J. Psychol.* 12:97–102.

GOLD, P. E., W. FARRELL, and R. A. KING, 1971. Retrograde amnesia after localized brain shock in passive-avoidance learning. *Physiol. Behav.* 7:709–712.

GROSSMAN, S. P., 1969. Facilitation of learning following localized intracranial injections of pentylenetetrazol. *Physiol. Behav.* 4:625–628.

GROSSMAN, S. P., and H. MOUNTFORD, 1964. Learning and extinction during chemically induced disturbance of hippocampal functions. *Amer. J. Physiol.* 207:1387–1393.

HALJAMÄE, H., and P. W. LANGE, 1972. Calcium content and conformational changes of S-100 protein in the hippocampus during training. *Brain Res* 38:131–142.

HAYWOOD, J., and S. P. R. ROSE, 1970. Effects of an imprinting procedure on RNA polymerase activity in the chick brain. *Nature (Lond.)* 228:373–374.

HERZ, M. J., and H. V. S. PEEKE, 1971. Impairment of extinction with caudate nucleus stimulation. *Brain Res.* 33:519–522.

HOKIN, L. E., 1969a. Phospholipid metabolism and functional activity of nerve cells. In *The Structure and Function of Nervous Tissue*, G. H. Bourne, ed. New York: Academic Press, pp. 161–184.

HOKIN, M. R., 1969b. Effect of norepinephrine on ^{32}P incorporation into individual phosphatides in slices from different areas of the guinea pig brain. *J. Neurochem.* 16:127–134.

HOKIN, L. E., 1965. Autoradiographic localization of the acetylcholine-stimulated synthesis of phosphatidylinositol in the superior cervical ganglion. *Proc. Nat. Acad. Sci. USA* 53:1369–1376.

HOKIN, L. E., 1966. Effects of acetylcholine on the incorporation of ^{32}P into various phospholipids in slices of normal and denervated superior cervical ganglia of the cat. *J. Neurochem.* 13:179–184.

HOKIN, M. R., L. E. HOKIN, and W. D. SHELP, 1960. The effects of acetylcholine on the turnover of phosphatidic acid and phosphoinositide in sympathetic ganglia, and in various parts of the central nervous system in vitro. *J. Gen. Physiol.* 44:217–226.

HORN, G., 1971. Biochemical, morphological and functional changes in the central nervous system associated with experience. *Activitas Nervosa Superior* 13:119–130.

HUDSPETH, W. J., and W. E. WILSONCROFT, 1969. Retrograde amnesia: Time dependent effects of rhinencephalic lesions. *J. Neurobiol.* 2:221–232.

HYDÉN, H., and P. W. LANGE, 1968. Protein synthesis in the hippocampal pyramidal cells of rats during a behavioral test. *Science* 159:1370–1373.

HYDÉN, H., and P. W. LANGE, 1970a. Protein changes in nerve cells related to learning and conditioning. In *The Neurosciences: Second Study Program*, F. O. Schmitt, ed. New York: The Rockefeller University Press, pp. 278–289.

HYDÉN, H., and P. W. LANGE, 1970b. Time sequence analysis of proteins in brain stem, limbic system and cortex during training. *Biochimica e Biologia Sperimentale* 9:275–285.

HYDÉN, H., and P. W. LANGE, 1972a. Protein changes in different brain areas as a function of intermittent training. *Proc. Nat. Acad. Sci. USA* 69:1980–1984.

HYDÉN, H., and P. W. LANGE, 1972b. Protein synthesis in hippocampal nerve cells during re-reversal of handedness in rats. *Brain Res.* 45:035–038.

ITOH, T., and J. H. QUASTEL, 1969. Ribonucleic acid biosynthesis in adult and infant rat brain in vitro. *Science* 164:79–80.

JARVIK, M. E., 1972. Effects of chemical and physical treatments on learning and memory. *Ann. Rev. Psychol.* 23:457–486.

JOHN, R. E., 1967. *Brain Mechanisms and Memory.* New York: Academic Press.

JONES, C. T., and P. BANKS, 1970a. The effect of electrical stimulation and ouabain on the uptake and efflux of L-[U-^{14}C]valine in chopped tissue from guinea-pig cerebral cortex. *Biochem. J.* 118:801–812.

JONES, C. T., and P. BANKS, 1970b. The effect of electrical stimulation on the incorporation of L-[U-^{14}C]valine into the protein of chopped tissue from guinea-pig cerebral cortex. *Biochem. J.* 118:791–800.

JONES, D. A., 1972. The relationship between amino acid incorporation into protein in isolated neocortex slices and the tissue content of free amino acid. *J. Neurochem.* 19:779–790.

JONES, D. A., and H. McILWAIN, 1971. Amino acid distribution and incorporation into proteins in isolated, electrically-stimulated cerebral tissues. *J. Neurochem.* 18:41–58.

JONES, D. A., and R. RODNIGHT, 1971. Protein-bound phosphorylserine in acid hydrolysates of brain tissue. The determination of [^{32}P]phosphorylserine by ion-exchange chromatography and electrophoresis. *Biochem. J.* 121:597–600.

KAKIUCHI, S., and T. W. RALL, 1968a. The influence of chemical agents on the accumulation of adenosine 3′,5′-phosphate in slices of rabbit cerebellum. *Mol. Pharmacol.* 4:367–378.

KAKIUCHI, S., and T. W. RALL, 1968b. Studies on adenosine 3′,5′-phosphate in rabbit cerebral cortex. *Mol. Pharmacol.* 4:379–388.

KERKUT, G. A., G. W. O. OLIVER, J. T. RICK, and R. J. WALKER, 1970. The effects of drugs on learning in a simple preparation. *Comp. Gen. Pharmac.* 1:437–483.

KERNELL, D., and R. P. PETERSON, 1970. The effect of spike activity versus synaptic activation on the metabolism of ribonucleic acid in a molluscan giant neuron. *J. Neurochem.* 17:1087–1094.

KESNER, R. P., and H. S. CONNER, 1972. Independence of short- and long-term memory: A neural system analysis. *Science* 176:432–434.

KESNER, R. P., and R. W. DOTY, 1968. Amnesia produced in cats by local seizure activity initiated from the amygdala. *Exp. Neurology* 21:58–68.

KLAINER, L. M., Y.-M. CHI, S. L. FREIDBERG, T. W. RALL, and E. W. SUTHERLAND, 1962. Adenyl cyclase. IV. The effects of neurohormones on the formation of adenosine 3′,5′-phosphate by preparations from brain and other tissues. *J. Biol. Chem.* 237:1239–1243.

KOHEN, E., and C. KOHEN, 1966. A study of mitochondrial-extramitochondrial interactions in giant tissue culture cells by microfluorimetry-microelectrophoresis. *Histochemie* 7:339–347.

KUO, J.-F., T.-P. LEE, P. L. REYES, K. G. WALTON, T. E. DONNELLY, and P. GREENGARD, 1972. Cyclic nucleotide-dependent protein kinases. X. An assay method for the measurement of guanosine 3′,5′-monophosphate in various biological materials and a study of agents regulating its levels in heart and brain. *J. Biol. Chem.* 247:16–22.

LANDE, S., J. B. FLEXNER, and L. B. FLEXNER, 1972. Effect of corticotropin and desglycinamide^9-lysine vasopressin on suppression of memory by puromycin. *Proc. Nat. Acad. Sci. USA* 69:558–560.

LARRABEE, M. G., 1968. Transynaptic stimulation of phosphatidylinositol metabolism in sympathetic neurons in situ. *J. Neurochem.* 15:803–808.

LARRABEE, M. G., and W. S. LEICHT, 1965. Metabolism of phosphatidyl inositol and other lipids in active neurons of sympathetic ganglia and other peripheral nervous tissues. The site of the inositide effect. *J. Neurochem.* 12:1–13.

LIDSKY, A., and B. M. SLOTNICK, 1970. Electrical stimulation of the hippocampus and electroconvulsive shock produce similar amnestic effects in mice. *Neuropsychologia* 8:363–369.

LLINAS, R., J. R. BLINKS, and C. NICHOLSON, 1972. Calcium transient in presynaptic terminal of squid giant synapse: Detection with aequorin. *Science* 176:1127–1129.

McAFFEE, D. A., M. SCHORDERET, and P. GREENGARD, 1971. Adenosine 3′,5′-monophosphate in nervous tissue: Increase associated with synaptic transmission. *Science* 171:1156–1158.

McBRIDE, W. J., and J. D. KLINGMAN, 1972. The effects of electrical stimulation and ionic alterations on the metabolism of amino acids and proteins in excised superior cervical ganglia of the rat. *J. Neurochem.* 19:865–880.

MACHLUS, B. J., J. E. WILSON, and E. GLASSMAN, 1973a. Brain phosphoproteins. I. The effect of short experiences on the incorporation of radioactive phosphate into nuclear protein of rat brain, (in preparation).

MACHLUS, B. J., J. E. WILSON, and E. GLASSMAN, 1973b. Brain phosphoproteins. II. The effect of various behaviors and reminding experiences on the incorporation of radioactive phosphate, (in preparation).

MacINNES, J. W., E. H. McCONKEY, and K. SCHLESINGER, 1970. Changes in brain polyribosomes following an electroconvulsive seizure. *J. Neurochem.* 17:457–460.

MAHUT, H., 1962. Effects of subcortical electrical stimulation on learning in the rat. *J. Comp. Physiol. Psychol.* 55:472–477.

MAITRA, P. K., A. GHOSH, B. SCHOENER, and B. CHANCE, 1964. Transients in glycolytic metabolism following electrical activity in electrophorus. *Biochim. Biophys. Acta* 88:112–119.

MARK, R. F., and M. E. WATTS, 1971. Drug inhibition of memory formation in chickens. I. Long-term memory. *Proc. R. Soc. Lond. (Biol.)* 178:439–454.

MOORE, W. V., and D. B. WETLAUFER, 1971. The circular dichroism of synaptosomal membranes: Studies of interactions with neuropharmacological agents. *J. Neurochem.* 18 1167–1178.

MULLONEY, B., and A. SELVERSTON, 1972. Antidromic action potentials fail to demonstrate known interactions between neurons. *Science* 177:69–72.

ORREGO, F., 1967. Synthesis of RNA in normal and electrically stimulated brain cortex slices in vitro. *J. Neurochem.* 14:851–858.

ORREGO, F., and F. LIPMANN, 1967. Protein synthesis in brain slices. Effects of electrical stimulation and acidic amino acids. *J. Biol. Chem.* 242:665–671.

PAOLINO, R. M., and H. M. LEVY, 1971. Amnesia produced by spreading depression and ECS: Evidence for time-dependent memory trace localization. *Science* 172:746–749.

PEEKE, H. V. S., and M. J. HERZ, 1971. Caudate nucleus stimulation retroactively impairs complex maze learning in the rat. *Science* 173:80–82.

PRIVES, C., and J. H. QUASTEL, 1969. Effect of cerebral stimulation on biosynthesis of nucleotides and RNA in brain slices in vitro. *Biochim. Biophys. Acta* 182:285–294.

QUARTERMAIN, D., and B. S. McEWEN, 1970. Temporal characteristics of amnesia induced by protein synthesis inhibitor: Determination by shock level. *Nature (Lond.)* 228: 677–678.

QUEVEDO, J. C., P. G. BOSQUE, M. L. ANDRÉS, and M. C. C. GARCIA, 1971. Cytoplasmic RNA in the rat hippocampus after a learning experience: An autoradiographic study. *Acta Anat.* 79:360–366.

RAHMANN, H., 1973. Radioactive studies of changes in protein metabolism by adequate and inadequate stimulation in the optic tectum of teleostes. In *Memory and Transfer of Information*, H. P. Zippel, ed. New York: Plenum Publishing Corporation (in press).

RANDT, C. T., D. QUARTERMAIN, M. GOLDSTEIN, and B. ANAGNOSTE, 1971. Norepinephrine biosynthesis inhibition: Effects on memory in mice. *Science* 172:498–499.

ROBERTS, E., and S. MATTHYSSE, 1970. Neurochemistry: At the crossroads of neurobiology. *Ann. Rev. Biochem.* 39:777–820.

ROSE, S. P. R., 1970. Neurochemical correlates of learning and environmental change. In *Short-term Changes in Neural Activity and Behaviour*, G. Horn and R. A. Hinde, eds. Cambridge, England: Cambridge University Press, pp. 517–550.

Rose, S. P. R., 1972. Changes in amino acid pools in the rat brain following first exposure to light. *Brain Res.* 38:171–178.

Schwartz, J. H., V. F. Castellucci, and E. R. Kandel, 1971. Functioning of identified neurons and synapses in abdominal ganglion of *Aplysia* in absence of protein synthesis. *J. Neurophysiol.* 34:939–953.

Serota, R. G., 1971. Acetoxycycloheximide and transient amnesia in the rat. *Proc. Nat. Acad. Sci. USA* 68:1249–1250.

Serota, R. G., R. B. Roberts, and L. B. Flexner, 1972. Acetoxycycloheximide-induced transient amnesia: Protective effects of adrenergic stimulants. *Proc. Nat. Acad. Sci. USA* 69:340–342.

Shimizu, H., C. R. Creveling, and J. W. Daly, 1970. Cyclic adenosine 3′,5′-monophosphate formation in brain slices: Stimulation by batrachotoxin, ouabain, veratridine, and potassium ions. *Mol. Pharmacol.* 6:184–188.

Skrzipek, K-H., 1969. Die proteinsynthese des tectum opticum in Abhängigkeit von der Gestalt intermittierender Lichtmuster bei *Carassius carassius* L. (Pisces). *J. Hirnforschung* 77:414–416.

Sokoloff, L., 1961. Local cerebral circulation at rest and during altered cerebral activity induced by anesthesia or visual stimulation. In *Regional Neurochemistry*, S. S. Kety and J. Elkes, eds. London: Pergamon Press, pp. 107–117.

Squire, L. R., and S. H. Barondes, 1972. Variable decay of memory and its recovery in cycloheximide-treated mice. *Proc. Nat. Acad. Sci. USA* 69:1416–1420.

Stein, D. G., and S. L. Chorover, 1968. Effects of posttrial electrical stimulation of hippocampus and caudate nucleus on maze learning in the rat. *Physiol. Behav.* 3:787–791.

Stoller, W., and M. J. Wayner, 1968. Lack of effect of neural activity on RNA synthesis in rat dorsal root ganglion cells. *Physiol. Behav.* 3:941–945.

Tasaki, I., L. Carnay, and A. Watanabe, 1969. Transient changes in extrinsic fluorescence of nerve produced by electric stimulation. *Proc. Nat. Acad. Sci. USA* 64:1362–1368.

Tasaki, I., A. Watanabe, and M. Hallett, 1972. Fluorescence of squid axon membrane labelled with hydrophobic probes. *J. Membrane Biol.* 8:109–132.

Thompson, R., 1958. The effect of intracranial stimulation on memory in cats. *J. Comp. Physiol. Psychol.* 51:421–426.

Ungar, G., 1972. Molecular approaches to neural coding. *Int. J. Neurosci.* 3:193–200.

Uphouse, L. L., J. W. MacInnes, and K. Schlesinger, 1972a. Effects of conditioned avoidance training on polyribosomes of mouse brain. *Physiol. Behav.* 8:1013–1018.

Uphouse, L. L., J. W. MacInnes, and K. Schlesinger, 1972b. Uridine incorporation into polyribosomes of mouse brain after escape training in an electrified T-maze. *Physiol. Behav.* 8:1019–1023.

Vardaris, R. M., and K. E. Schwartz, 1971. Retrograde amnesia for passive avoidance produced by stimulation of dorsal hippocampus. *Physiol. Behav.* 6:131–135.

Watson, W. E., 1965. An autoradiographic study of the incorporation of nucleic-acid precursors by neurons and glia during nerve stimulation. *J. Physiol.* 180:754–765.

Watts, M. E., and R. F. Mark, 1971. Separate actions of ouabain and cycloheximide on memory. *Brain Res.* 25:420–423.

Wegener, G., 1970. Enhancement of protein synthesis in the optic tectum of frogs following light stimulation. An autoradiographic investigation. *Exp. Brain Res.* 10:363–379.

Wilson, D. L., 1971. Molecular weight distribution of proteins synthesized in single, identified neurons of *Aplysia. J. Gen. Physiol.* 57:26–40.

Wilson, D. L., and R. W. Berry, 1972. The effect of synaptic stimulation on RNA and protein metabolism in the R2 soma of *Aplysia. J. Neurobiol.* 3:369–379.

Woodson, P. B. J., W. T. Schlapfer, and S. H. Barondes, 1972. Postural avoidance learning in the headless cockroach without detectable changes in ganglionic cholinesterase. *Brain Res.* 37:348–352.

Wyers, E. J., and S. A. Deadwyler, 1971. Duration and nature of retrograde amnesia produced by stimulation of caudate nucleus. *Physiol. Behav.* 6:97–103.

Wyers, E. J., H. V. S. Peeke, J. S. Williston, and M. J. Herz, 1968. Retroactive impairment of passive avoidance learning by stimulation of the caudate nucleus. *Exp. Neurol.* 22:350–366.

Zemp, J. W., J. E. Wilson, K. Schlesinger, W. O. Boggan, and E. Glassman, 1966. Brain function and macromolecules. I. Incorporation of uridine into RNA of mouse brain during short-term training experience. *Proc. Nat. Acad. Sci. USA* 55:1423–1431.

Zornetzer, S. F., 1972. Brain stimulation and retrograde amnesia in rats: A neuroanatomical approach. *Physiol. Behav.* 8:239–244.

Zornetzer, S. F., and J. L. McGaugh, 1972. Electrophysiological correlates of frontal cortex-induced retrograde amnesia in rats. *Physiol. Behav.* 8:233–238.

58 Biochemical Correlates of Brief Behavioral Experiences

A. DUNN, D. ENTINGH, T. ENTINGH,
W. H. GISPEN, B. MACHLUS,
R. PERUMAL, H. D. REES, and
L. BROGAN

ABSTRACT Biochemical changes associated with conditioned avoidance training in mice and rats have been investigated. Some of these changes are related to the training and some are not. The biochemical changes include phosphorylation of nuclear protein, incorporation of phosphate into synaptosomal protein, incorporation of glucosamine into gangliosides, incorporation of amino acids into protein, and incorporation of fucose into glycoprotein. The data indicate that brain metabolism responds in a variety of ways to brief environmental stimulation. Moreover, the chemical responses exhibit differing specificities with respect to the behavioral, anatomical, and temporal aspects of the stimulation, which may indicate that they are related to different aspects of the experience.

MEMBERS OF THE Division of Chemical Neurobiology of the University of North Carolina are studying biochemical changes that occur in the brains of mice and rats during and shortly after brief environmental stimulation or training procedures. Although the work is aimed at the eventual elucidation of biochemical events necessary for the formation of memory, effects of brief exposures to simple stimuli are also being studied. This broad approach has been adopted because learning is a complex process, and it is thought likely that particular components of the training situation, such as stress, arousal, attention, motivation, etc., although often considered to be extraneous to memory storage itself, may in fact be necessary aspects of the learning process. Similarly, since behavior is in general produced by interactions between the brain, the body, and its environment, metabolic responses of peripheral organs to environmental stimulation are being studied concurrently to provide clues for understanding the changes that occur in the brain.

Mice and rats were trained in an active avoidance task using footshock as an aversive stimulus. C57Bl/6J mice (♂, 6–8 weeks old; Jackson Laboratories) were trained for

A. DUNN, D. ENTINGH, T. ENTINGH, W. H. GISPEN, B. MACHLUS, R. PERUMAL, H. D. REES, and L. BROGAN. Division of Chemical Neurobiology, Department of Biochemistry, School of Medicine, University of North Carolina, Chapel Hill, North Carolina

15 min in the jump box in which the mice learn to jump to a safe shelf when a light and buzzer are activated in order to avoid an electric shock transmitted to the grid floor (Zemp et al., 1966; Wilson and Glassman, 1972). Yoked animals received the same stimuli in an adjacent compartment in the box but were unable to avoid the footshock. Quiet animals were left undisturbed after injection in individual cages in a separate room but in a normal laboratory environment. In some experiments mice were trained in an automated version of the above apparatus with essentially similar results. Rats were trained for 5 min in a step-up avoidance situation in which they learned to step onto a shelf to avoid a shock from the electrified grid floor (Coleman et al., 1971; Wilson and Glassman, 1972).

These behavioral tasks constitute a significant emotional experience in the lives of the animals, because they encounter novel stimuli that are quite unlike anything previously experienced. Thus the situation is probably very different from learning in the natural environment. The novelty of the experiences may explain the large magnitude and general anatomical distribution of some of the biochemical effects observed.

The metabolism of the animals has been studied primarily by the use of radioactive precursors of macromolecules. Macromolecules have been chosen for study because of their comparative stability and their structural and regulatory roles in the nervous system. In most cases a pool correction factor has been applied to derive the relative radioactivity (RR), defined as the radioactivity in the product macromolecule divided by that in a precursor molecule from the same tissue. This correction reduces the variability due to injection and partially corrects for variations in uptake of the precursor, but it should not be assumed in the absence of other data that changes in RR necessarily reflect changes in macromolecular synthesis. Changes in the time course of uptake or compartmentation phenomena may confound such a simplistic interpretation (see Glassman, this volume).

A. DUNN ET AL. 679

TABLE I

TABLE I

Behavior-related chemical changes

Chemical	$\dfrac{T}{Q}$	$\dfrac{T}{Y}$	$\dfrac{Y}{Q}$	$\dfrac{CC}{Q}$	$\dfrac{PT}{Q}$	$\dfrac{Ext}{Y}$	$\dfrac{Rem}{Q}$	Reference
Training-Related Responses								
RNA		+	0					Zemp et al., 1966
Polyribosomes†		+	0	0	0	+		Adair et al., 1968a, b Coleman et al., 1971
NAEP	+		0	0	+	+	+	Machlus, 1971
Synaptosomal phosphoprotein†	+	+	0		0	+		Perumal and Gispen, unpublished observations
Stimulus-Related Responses								
Protein	+	0	+		+*			Rees and Brogan, unpublished observations
Glycoprotein	+							Entingh and Entingh, unpublished observations
Uncharacterized								
Histones	—							Machlus, 1971
Gangliosides	+							Dunn and Hogan, unpublished observations

T: Trained; Q: Quiet; Y: Yoked; CC: Classically-conditioned; PT: Prior-Trained; Ext: Extinguished; Rem: Prior-trained Reminded.

*Increase smaller than in T/Q.

†This response is apparently related to *novel* training.

Significant differences between trained and quiet animals were observed in the radioactivity of isolated macromolecules with a number of different precursors when appropriate times of injection and sacrifice were chosen. However, there were several types of specificity in the metabolic changes. Changes were maximal at different times; they were manifested in different regions of the brain, and some were also present in organs other than the brain. Moreover, there was behavioral specificity. Yoked animals or shocked animals were biochemically similar either to quiet or to trained animals depending on the precursor. Thus the metabolic responses may, in the first instance, be classified as related to training or to general stimulation (Table I).

Phosphorylation of nuclear proteins

The phosphorylation of nuclear proteins in the rat was studied using a double isotope procedure injecting inorganic phosphate labeled with either ^{32}P or ^{33}P intracranially. Thirty minutes later one rat was trained for 5 min in the step-up avoidance situation and then immediately sacrificed. The brain was homogenized with that of a quiet rat that had been injected at the same time as the trained rat with the other phosphorus isotope. Isolated nuclei were extracted with acid and the proteins solubilized in this way separated into histone and nonhistone fractions by ion-exchange chromatography. Adenosine monophosphate (AMP) isolated from the

brain homogenate was used as the correction factor for the uptake of radioactive phosphate. The nonhistone acid-extractable nuclear protein (NAEP) from the brain of the trained rat was found to contain approximately twice as much radioactive phosphorus as that from the quiet rat (Table II). In contrast, the pooled histones from the trained rat contained only about 60% of the radioactive

TABLE II

Incorporation of ^{32}P or ^{33}P into rat brain NAEP during one-way step-up avoidance

Treatment	n	NAEP/AMP† Increase (%)
Trained/quiet	30	$+104 \pm 11$*
Yoked/quiet	20	-5 ± 59
Shocked/quiet	4	-11 ± 14
6-day prior-trained performing/quiet	8	$+95 \pm 4$*
6-day prior-trained reminded/quiet	8	$+78 \pm 14$*
6-day prior-trained handled/quiet	4	$+84 \pm 6$*

*$P < 0.01$ (2-tailed t-test).

†Mean $\pm$ standard deviation in this and all subsequent tables.

A dual-isotope method was used in which rats were injected intracranially with 100 μc [^{32}P]H$_3$PO$_4$ or [^{33}P]H$_3$PO$_4$. Thirty minutes later they were trained for 5 min and sacrificed immediately. Each brain was homogenized with the brain of a quiet rat that had received the other radioisotope. NAEP and AMP were isolated from the homogenate and their radioactivities determined. (Data from Machlus, 1971.)

phosphorus of those from the quiet rat (Machlus, 1971).

The time course of incorporation of radioactivity in quiet rats showed that after 10 min the radioactivity in both AMP and NAEP was approximately constant for up to 1 hr. The pronounced increase in labeling of the NAEP was apparently a rapid response to the training. When the training was prolonged for a further 5 min or when the animal was left quiet for 5 min after training before sacrifice, the increase in radioactivity in NAEP was not observed (Figure 1). Analysis by gross dissection indicated that the effect was confined to the basal forebrain.

Studies with degradative enzymes suggested that the phosphate was covalently attached to the protein. Moreover, the radioactivity in phosphoserine isolated following hydrolysis of the NAEP was increased after training. Even more significantly, the molar ratio of phosphoserine to serine in the NAEP increased in trained compared with quiet rats, *whether or not* they had been injected with radioactive precursor (Machlus, 1971). This result is consistent with the time course data and is important, because it indicates that the effect was not due to permeability or other changes affecting the radioactive precursors.

The increase of radioactive phosphate in NAEP was not observed in rats that had been yoked to the trained animal or had received random electric footshocks. However, the increase was observed in trained rats that had

been prior-trained for the 6 preceding days (Table II). Indeed, if on the seventh day the rat was not trained but merely placed in the conditioning apparatus (reminded), or merely handled 30 min after the injection without being placed in the apparatus, the radioactivity in NAEP was also increased (Table II). This response was specific for prior-trained rats and was not observed in animals that

TABLE III

Incorporation of ^{32}P or ^{33}P into mouse brain NAEP during training

Treatment	n	NAEP/AMP Increase (%)
1. Step-up avoidance		
Trained/quiet	4	+67 ±5**
2. Jump-box avoidance		
Trained/quiet	4	+52 ±16*
Classically-conditioned/quiet	4	−6 ±15
6-day prior-trained performing/quiet	4	+87 ±4**
1-day prior-trained reminded/quiet	8	+39 ±20**

*P < 0.02.
**P < 0.01 (2-tailed *t*-test).

Mice were injected intracranially with 33 μc [^{32}P]H$_3$PO$_4$ or [^{33}P]H$_3$PO$_4$. Thirty minutes later they were trained for 5 min and then immediately sacrificed and NAEP and AMP isolated from the brain homogenate. In classically-conditioned animals the shock followed the light and buzzer, but mice were unable to avoid the shock. (Data from Machlus, 1971.)

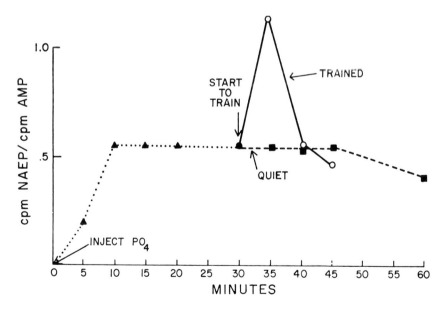

FIGURE 1 Time course of the incorporation of radioactive phosphate into nonhistone acid-extractable proteins (NAEP) of rat brain after intracranial injection. Rats were killed at various times after injection and NAEP and AMP isolated from the brain homogenate and their radioactivity determined. *Dotted line:* Quiet rats. *Solid line:* Rats trained 30 min after injection. For training details see text and the caption to Table II. (Data from Machlus, 1971.)

had been shocked instead of trained for 6 days and were treated similarly on the seventh day. This suggests that *after* training the response may be linked specifically with stimuli involved in the training, including even the handling involved in placing the animal on the grid floor of the training apparatus at the start of each training trial.

Mice trained for 5 min either in the step-up avoidance or the jump box showed a similar but somewhat smaller increase in labeling of NAEP with phosphate (Table III). In mice too, the response occurred in prior-trained animals but was not observed after a series of unavoidable footshocks.

Incorporation of phosphate into synaptosomal protein

The phosphorylation of synaptosomal proteins in mouse brain has been studied after 15 min of jump-box training. Using a double isotope procedure similar to that described above for NAEP, an increase of about 30% in the radioactive phosphate associated with synaptosomal proteins was observed in trained compared with quiet mice (Table IV). The effect was not observed in yoked mice. This increase in protein-bound radioactivity was associated with the membrane fraction of osmotically lysed synaptosomes. As in the case of NAEP, the increase occurred in the phosphoserine isolated following hydrolysis of the protein (Perumal and Gispen, unpublished observations).

Incorporation of glucosamine into gangliosides

Gangliosides are membrane constituents thought to be present at high concentrations at or near synapses. Their

TABLE IV

Incorporation of radioactive phosphate into mouse brain synaptosomal protein

Treatment	n	Synaptosomal Protein/AMP Increase (%)
Trained/quiet	21	$+32 \pm 30$*
Yoked/quiet	12	$+1 \pm 18$

*$P < 0.005$ (2-tailed t-test).

Mice were injected intracranially with 20 μc [^{32}P]H$_3$PO$_4$ or [^{33}P]H$_3$PO$_4$. Thirty minutes later they were trained for 15 min in the jump-box and sacrificed immediately. Synaptosomes were isolated and radioactivies in AMP and synaptosomal phosphoprotein determined. (Perumal and Gispen, unpublished observations.)

metabolism was studied by observing the incorporation of D[1-^3H]glucosamine. Training did not change the incorporation of ^3H into the residue left after lipid extraction (radioactivity mainly in glycoproteins, or into nonganglioside lipid fractions), but did consistently increase the incorporation of ^3H into the total ganglioside fraction by about 25% (Table V).

Incorporation of amino acids into protein

Changes in brain protein metabolism were studied by monitoring the incorporation of radioactive amino acids following jump-box training of mice. When [^3H]lysine was injected subcutaneously into mice 20 min after 15 min of jump-box training, the total ^3H content of the brain was significantly increased by 13%, 10 min later. When

TABLE V

Incorporation of D[1-^3H]glucosamine into mouse brain

Treatment	C-M Soluble	Nonlipid Residue	Lipids	Gangliosides
		Dpm $\times 10^{-3}$		
Trained	45.4 ± 9.3	255 ± 43	3.90 ± 0.40	2.84 ± 0.47
Yoked	43.3 ± 2.0	251 ± 42	3.94 ± 0.33	2.35 ± 0.47
Quiet	52.2 ± 13.3	248 ± 41	4.15 ± 0.63	2.26 ± 0.38
		Ratio		
Trained/quiet	0.87	1.03	0.94	1.26*
Trained/yoked	1.05	1.02	0.99	1.21
Yoked/quiet	0.83	1.01	0.95	1.04

*$P < 0.025$ (t-test),
$n = 6$ for each group.

Mice were injected subcutaneously with 25 μc D[1-^3H]glucosamine 15 min before 15 min jump-box training. They were sacrificed 30 min later and gangliosides extracted as described by Suzuki (1965). C-M soluble: Chloroform-methanol soluble (Folch extract); Nonlipid residue: Residue after Folch extraction (protein and glycoprotein); Lipids: Lower, hydrophobic phase (nonganglioside lipid); Gangliosides: Dialyzed upper phase (monitored by thin layer chromatography on silica gel). (Dunn and Hogan, unpublished observations.)

the RSA was calculated to correct for the increased uptake, there was still a statistically significant increase of ^{3}H incorporation into protein in trained animals compared with quiet (Table VI). The RR after the 10 min pulse was elevated to approximately the same extent when animals were sacrificed 15, 30, or 45 min after training. Similar increases in RR were observed in trained mice using [4,5-^{3}H]leucine or [1-^{14}C]leucine as precursors of protein.

TABLE VI

Incorporation of [4,5-^{3}H]lysine into mouse brain protein

Treatment	n	Mean RR	% Increase over Quiet
Quiet	52	0.127 ± 0.014	[0]
Trained	18	0.148 ± 0.014	+17**
Yoked	4	0.144 ± 0.016	+13*
30 Buzzers	13	0.146 ± 0.018	+15**
20 Shocks	14	0.143 ± 0.015	+13**
30 Lights	6	0.132 ± 0.009	+4

*P < 0.05; **P < 0.001 (2-tailed *t*-test).
Mice were trained or otherwise stimulated for 15 min in the jump box. Quiet animals were undisturbed prior to injection. Twenty minutes later they were injected subcutaneously with 30 μc [4,5-^{3}H]lysine and sacrificed 10 mins later. RR = dpm in dried trichloroacetic acid precipitate/dpm in dried trichloroacetic acid supernatant. Independent experiments have shown that more than 95% of the ^{3}H in the dried trichloroacetic acid supernatant was in lysine. (Rees and Brogan, unpublished observations.)

The effect, however, was not specific to training, and an effect of similar magnitude was observed in yoked mice (Table VI). When the various components of the training situation were tested separately, buzzers or shocks, but not lights, elicited the increased RR (Table VI). The effect may thus be regarded as stimulus-related rather than training-related.

The increased RR occurred generally throughout the brain (dissected into 6 parts) and was significant everywhere but in the dorsal cortex and thalamus-hypothalamus. The liver also showed an RR increase of greater magnitude than the brain under the same behavioral conditions. This liver response was maximal immediately after training and disappeared within 1 hr. The elevation of RR with stimulation was also observed in the brains and the livers of rats and of adrenalectomized mice.

Incorporation of fucose into glycoproteins

The metabolism of fucose-containing glycoproteins in brain was studied by following the incorporation of L[1-^{3}H]fucose. Jump-box training of mice produced an increase of the incorporation of radioactivity into glycoprotein of about 20–30%, relative to quiet mice. Yoked mice also showed this change suggesting that it was not specific to the training aspects of the experience. Maximal effects were observed when fucose was injected shortly before the training and the animals sacrificed very soon after its completion (Entingh and Entingh, unpublished observations).

Conclusions

Taken together, these data indicate that diverse metabolic changes occur in the brains of mice and rats undergoing avoidance training. The characteristics of these various responses are dissimilar, and they are thus not due to a single general disturbance of cerebral metabolism. These biochemical responses are summarized in Table I, where they have been divided into two categories, *training-related* and *stimulus-related*. Training-related responses are observed in trained but not in yoked animals. Stimulus-related responses appear in both trained and yoked animals. In addition two types of training-related responses may be differentiated. The change in the phosphorylation of NAEP occurs only in animals that have been trained but may be induced by subsequent related experiences. In contrast the increased radioactivity in polyribosomes after administration of radioactive uridine (Adair et al., 1968a, 1968b) does not occur when prior-trained mice perform the conditioned avoidance. It does, however, occur when prior-trained mice extinguish their avoidance behavior (Coleman et al., 1971). Thus the effect appears to be related to *novel* training. There are insufficient behavioral data to classify some of the metabolic responses, and other categories may exist.

Anatomically the effects apparently occurred in large regions of the brain, but there was regional specificity. Most of the increase of uridine incorporation into RNA was confined to the subcortical forebrain (Zemp et al., 1967), as was the increased incorporation of phosphate into NAEP; whereas the lysine incorporation response was more diffuse, with maximal effects in ventral cortex, basal ganglia, septum and hippocampus, and cerebellum and brainstem. Bowman and Strobel (1969) and Bateson et al. (1972) have also observed anatomical specificity in their behavior-related biochemical changes in rats and chicks respectively.

Specificity is also apparent in the time relationships between behavioral and metabolic events. The change in the incorporation of [^{3}H]fucose into glycoprotein was maximal when the pulse was confined to the training period itself and the changes in the incorporation of uridine into RNA and polyribosomes, and the phosphorylation of

NAEP were largest when the animals were sacrificed immediately after training. However, the increase in [^{3}H]lysine incorporation occurred after training or stimulation was completed.

Further work is needed to characterize biochemically the responses to environmental stimulation. Although measurements with radioisotopes provide ambiguous data, since observed differences in radioactivity may be due to changes in blood flow or uptake, the changes, whatever their nature, are behaviorally and anatomically specific. At present, only one biochemical change, the increase in phosphoserine content of NAEP, has been detected without the use of radioactive precursors. However, the isotope data will give clues to the occurrence of metabolic changes, which may then be amenable to investigation by other means. Characterization of these changes may then suggest their immediate cause and possible biochemical consequences. In this way a clear understanding of the biochemical responses of the brain to environmental stimulation may be generated. It is hoped that this will eventually lead to the elucidation of the molecular events underlying the plasticity of the brain.

ACKNOWLEDGMENTS The contributions of Drs. E. Glassman, J. E. Wilson, and E. Hogan in many fruitful discussions are gratefully acknowledged. This research was supported by grants from the U.S. Public Health Service (MH 18136, NS 07457); the U.S. National Science Foundation (GB 18551); the Ciba-Geigy Corporation; and a Faculty Grant (AF 339) from the School of Medicine, University of North Carolina.

REFERENCES

ADAIR, L. B., J. E. WILSON, J. W. ZEMP, and E. GLASSMAN, 1968a. Brain function and macromolecules. III. Uridine incorporation into polysomes of mouse brain during short-term avoidance conditioning. *Proc. Nat. Acad. Sci. USA* 61: 606–613.

ADAIR, L. B., J. E. WILSON, and E. GLASSMAN, 1968b. Brain function and macromolecules. IV. Uridine incorporation into polysomes of mouse brain during different behavioral experiences. *Proc. Nat. Acad. Sci. USA* 61:917–922.

BATESON, P. P. G., G. HORN, and S. P. R. ROSE, 1972. Effects of early experience on regional incorporation of precursors into RNA and protein in the chick brain. *Brain Res.* 39:449–465.

BOWMAN, R. E., and D. A. STROBEL, 1969. Brain RNA metabolism in the rat during learning. *J. Comp. Physiol. Psychol.* 67:448–456.

COLEMAN, M. S., J. E. WILSON, and E. GLASSMAN, 1971. Incorporation of uridine into polysomes of mouse brain during extinction. *Nature (Lond.)* 229:54–55.

MACHLUS, B., 1971. Phosphorylation of nuclear proteins during behavior of rats. Ph.D. thesis, University of North Carolina, Chapel Hill, North Carolina.

SUZUKI, K., 1965. The pattern of mammalian brain gangliosides II. Evaluation of the extraction procedures, post-mortem changes and the effect of formalin preservation. *J. Neurochem.* 15:81–85.

WILSON, J. E., and E. GLASSMAN, 1972. The effect of short-term training experiences on the incorporation of radioactive precursors into RNA and polysomes of brain. In *Methods of Neurochemistry*, Vol. II, R. Fried, ed. New York: Marcel Dekker, pp. 53–72.

ZEMP, J. W., J. E. WILSON, K. SCHLESINGER, W. O. BOGGAN, and E. GLASSMAN, 1966. Brain function and macromolecules, I. Incorporation of uridine into RNA of mouse brain during a short-term behavioral experience. *Proc. Nat. Acad. Sci. USA* 55:1423–1431.

ZEMP, J. W., J. E. WILSON, and E. GLASSMAN, 1967. Brain function and macromolecules, II. Site of increased labeling of RNA of brains of mice during a short-term training experience *Proc. Nat. Acad. Sci. USA* 58:1120–1125.

59 Nutrition and the Brain

R. J. WURTMAN and J. D. FERNSTROM

ABSTRACT   The relationship of nutritional status, especially as regards protein ingestion, to mental function is explored. Neurochemical mechanisms whereby alterations in nutrition might affect mentation are surveyed. As a paradigm, the regulation of brain monoamine synthesis, especially serotonin, by blood and brain tryptophan is presented.

Introduction

IF THE ULTIMATE task of genetic mechanisms is to endow the cell with enzymes, the job of nutrition is to provide the substrates. These two tasks are not completely dissociated; enzymes are made from, induced by, and activated by nutritional inputs. However, a rough analogy can still be made: Genetics is to enzymes as nutrition is to substrates.

Considered in this perspective, the topic *Nutrition and the Brain* encompasses vastly more than the evidence that fish is or is not "brain food," or that macrobiotic diets do or do not predispose to transcendental insights or alpha rhythms: It includes all of those behavioral and biochemical processes that provide brain cells with compounds that they need but are unable to make for themselves.

Between the performance of the primary nutritional act—the eating or drinking or inhalation of the needed compound or its precursor—and the time that the nutrient interacts with the brain enzyme, a host of metabolic processes may intervene (Figure 1). For example, if the ingested material is a polymer (such as starch or protein), or part of a large, heterogeneous molecule (such as the trace metals bound to enzymes), it must first be digested, or dissociated into subunits, before it can be absorbed. Once it enters the circulation, the nutrient must be transported to the brain. It can do this dissolved in the plasma (e.g., glucose) or sequestered within erythrocytes (e.g., oxygen), or bound to proteins (e.g., fatty acids). En route, it may temporarily be converted to an insoluble storage form and retained within a reservoir of a tissue (e.g., the iron bound to ferritin in the liver) or within an extracellular space (e.g., the calcium in bone); or its chemical structure may be modified in the liver or elsewhere (e.g., the conversion of iodine to thyroxine in the thyroid gland). In addition, the level, or concentration, of the nutrient within the circulation may be maintained within a relatively narrow range through the operation of a *regulatory system* (e.g., glucose, calcium), or may simply reflect the vector sum of the various processes that are adding the nutrient to, and removing it from, the blood stream (e.g., amino acids, vitamin A). The ultimate entry of the nutrient into brain cells may involve simple diffusion, or be mediated by a specific transport system whose activity may be inhibited by competing substrates (e.g., related amino acids in the plasma), or stimulated by hormones or neurotransmitters.

Most of the evidence that nutritional state affects the brain has been derived from studies in which a particular nutrient was *chronically* underrepresented in the diet (e.g., thiamine in Wernicke's encephalopathy), or in which amounts of the nutrient that the test subject happened to require were unusually high (e.g., protein during the stage of brain development characterized by neuronal hyperplasia). This direction in nutrition research probably reflected the recognition that organisms contain substantial reservoirs for most of the nutrients that brain and other tissues require, and from the need to develop treatments for clinical syndromes (e.g., kwashiorkor, endemic goiter) resulting from human malnutrition. The absence of gross physiologic disturbances in subjects deprived of particular nutrients (e.g., glucose) for a few days probably reinforced this research strategy.

Such long-term studies have indeed demonstrated that chronic malnutrition can cause major and sometimes irreversible changes in brain weight, myelination, and even catecholamine content. However, more recent observations, summarized in this report, indicate that the brain can *also* respond to more subtle, transient changes in nutritional state. Specifically, the rate at which brain neurons synthesize serotonin, a putative neurotransmitter, is controlled on an hour-to-hour basis by their concentration of tryptophan; this concentration, in turn, varies characteristically depending on the composition of the food that has most recently been eaten (hence behavior, specifically feeding behavior, can control brain biochemistry, as well as vice versa). Paradoxically, the

R. J. WURTMAN and J. D. FERNSTROM   Laboratory of Neuroendocrine Regulation, Department of Nutrition and Food Science, Massachusetts Institute of Technology, Cambridge, Massachusetts

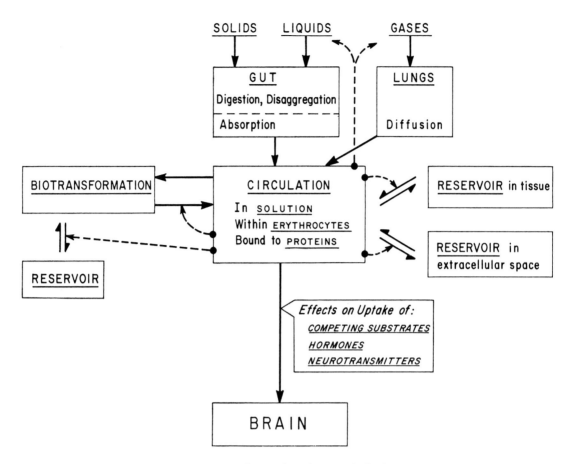

FIGURE 1   Delivery of nutrients to the brain.

quantities of free tryptophan available to the brain bear no simple relationship to the total amount of this amino acid present in the rest of the organism: As described below, protein-free diets can, acutely, cause greater elevations in brain tryptophan than diets containing protein. Brain nutrition need not parallel the general nutritional state.

Until fairly recently, it was widely held that the enormous importance of the brain conferred upon it a unique metabolic prerogative: The ability to take from the blood as much oxygen, glucose, or whatever else it needed, at rates that were more or less independent of the concentrations of these nutrients in the plasma. Under that formulation, a substrate or cofactor could become limiting for neurons only if it had to be synthesized within each neuron (e.g., S-adenosylmethionine; ATP) or if its supply in the body approached exhaustion. The clear dependence of brain serotonin upon plasma amino acid concentrations renders this earlier view obsolete. The normal brain is at least as responsive to transient changes in its nutritional state as any other organ in the body.

Control of brain serotonin by food consumption

Several years ago, we observed that the concentrations of tryptophan and of most other amino acids in human plasma undergo characteristic and parallel fluctuations during each 24-hr period (Wurtman et al., 1968; Wurtman, 1970); tryptophan levels are lowest at 2:00 to 4:00 AM, and rise 50 to 80% to attain a plateau late in the morning or early afternoon. The amplitude of the rhythm exhibited by any particular amino acid tends to vary inversely with its availability in the body; concentrations of relatively scarce amino acids (e.g., methionine, cysteine) rise and fall by as much as twofold during each 24-hr period, while the more abundant amino acids (e.g., glycine, glutamate) shift only 10 to 30%. Similar rhythms in the plasma concentrations of tryptophan and other amino acids were also observed in rats and mice (Figure 2); however, peak tryptophan levels, 35 to 150% above those at the daily nadir, occurred about 8 to 10 hr later than the peak in humans, a phenomenon probably attributable to the rat's tendency to consume most of its food during the hours surrounding the onset of the daily period of darkness. The plasma amino acid rhythms in humans were not simply the result of the cyclic ingestion of dietary protein, inasmuch as they persisted among volunteers who ate essentially no protein for 2 weeks (Wurtman et al., 1968). However, they disappeared in subjects placed on a total fast (Marliss et al., 1970); this finding suggested that they were *not* truly circadian or of endogenous origin, and that the most important factor in their genesis was nutritional, i.e., the release of insulin, and possibly other hormones, in response to dietary carbohydrate. (Insulin would be expected to raise or lower the plasma concentrations of amino acids by controlling their flux into muscle and other intracellular compartments.)

The mere existence of plasma amino acid rhythms did

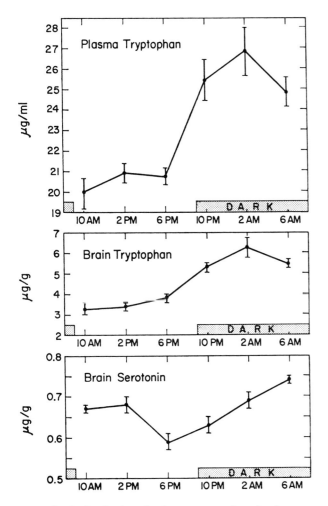

FIGURE 2 Daily rhythms in plasma tryptophan, brain tryptophan, and brain 5-HT. Groups of 10 rats kept in darkness from 9 PM to 9 AM were killed at intervals of 4 hr. Vertical bars indicate standard errors of the mean. (Reproduced from Wurtman and Fernstrom, 1972.)

not establish that such variations were of any consequence physiologically. To explore their possible significance, we set out to determine whether the naturally occurring daily fluctuations in the plasma concentration of a particular amino acid could actually influence its metabolic fate. This task could be accomplished by demonstrating that experimentally induced fluctuations of the same magnitude as those occurring diurnally caused parallel changes in the rate at which the amino acid is used for some purpose in the body (e.g., incorporation into proteins, conversion to a low molecular weight compound). The amino acid whose plasma concentration seemed most likely to influence its metabolic fate was tryptophan. The quantities of free and peptide-bound tryptophan present in the organism and in most foods were known to be the lowest of all the amino acids (Wurtman and Fernstrom, 1972); moreover, evidence has already been obtained that daily rhythms in the ingestion of tryptophan-containing proteins (and, presumably, in the concentration of tryptophan within portal venous blood) cause parallel rhythms in the aggregation of hepatic polysomes (Fishman et al., 1969), and in the activity and the synthesis of a specific liver protein, the enzyme tyrosine transaminase (Wurtman, 1970).

As the dependent variable in our study, we chose to look for possible changes in brain serotonin content in rats exhibiting a spontaneous daily rhythm in plasma tryptophan, or treated in such a way as to raise or lower plasma tryptophan. That the amount of tryptophan available to the brain might control serotonin synthesis was suggested by three lines of evidence: (1) the existence in rats and mice of diurnal rhythm in brain serotonin content (Figure 2) (Albrecht et al., 1956); (2) the unusually high Km for tryptophan shown by tryptophan hydroxylase, the enzyme that catalyzes the initial step in serotonin biosynthesis (Lovenberg et al., 1968); and (3) the repeated demonstrations that very high doses of tryptophan (i.e., 50 to 1600 mg/kg) could cause large increases in the brain concentrations of serotonin and its chief metabolite, 5-hydroxyindole acetic acid (5-HIAA) (Wurtman and Fernstrom, 1972).

Initial experiments were designed to determine whether brain serotonin could be increased by giving rats very low doses of tryptophan at a time of day (3 hr after the onset of the daily light period) when plasma and brain tryptophan, and brain serotonin concentrations are known to be low (Figure 1). It was hoped that a dose of tryptophan could be identified which (a) raised brain serotonin but (b) was smaller than the amount of the amino acid normally consumed by rats each day, and (c) did not elevate plasma and brain tryptophan levels beyond their normal daily peaks. All of these goals were met by administering 12.5 mg/kg i.p. of the amino acid to male rats weighing 150 to 200 g. This dose, constituting less than 5% of the amount of tryptophan that rats would be expected to ingest each day in 10 to 20 g of standard rat chow, produced peak elevations in plasma and brain tryptophan that were well within the ranges that occurred nocturnally in untreated animals, and caused brain serotonin levels to rise by 20 to 30% ($P < 0.01$) within 1 hr (Fernstrom and Wurtman, 1971a). Doses of 25 mg/kg caused elevations in both brain tryptophan and brain serotonin which were about double those observed in animals receiving the lower dose. Larger doses of tryptophan caused proportionate increases in brain tryptophan concentration but had little further effect on brain serotonin (Figure 3). Studies by others (Moir and Eccleston, 1968) have shown that further increments in tryptophan dose continue to produce increases in brain 5-HIAA, and thus presumably in serotonin biosynthesis. The failure of brain serotonin content to continue to rise when plasma and brain tryptophan are elevated beyond their normal dynamic ranges thus most likely reflects limitations in the ability of serotonin-producing neurons to store the amine.

The observed increase in brain serotonin caused by injecting very small doses of tryptophan was thought to be compatible with the hypothesis that the nocturnal rise in brain serotonin in normal rats is related to the daily rhythms in plasma and brain tryptophan (see Figure 2). It does not seem necessary to conclude that this substrate-induced rhythm in serotonin synthesis is the *only* factor responsible for the daily rhythm in brain serotonin content; for example, it is also possible that the serotonin rhythm reflects changes in the rates at which the mono-amine is released from neurons or metabolized intraneuronally.

Now that small increases in plasma tryptophan had been shown to cause parallel changes in brain serotonin, it became of interest to determine whether physiological decreases in the plasma amino acid also lowered the serotonin content. Hence we attempted to lower plasma tryptophan by giving rats insulin. It had not actually been shown that exogenous insulin lowers the plasma tryptophan concentration in rats, probably because a simple assay for tryptophan had become available only a few years previously. However, there was abundant evidence that insulin exerts this effect on almost all other amino acids examined, largely by enhancing their uptake into skeletal muscle (Wool, 1965).

Rats similar to those used in the previous experiments received a dose of insulin (2 units/kg i.p.) known to lower blood glucose levels. To our surprise, the hormone did not lower plasma tryptophan, but instead *increased* its concentration by 30 to 50% (Fernstrom and Wurtman, 1972a). This effect was independent of the route by which the insulin was administered; it was associated with a 55% fall in plasma glucose and with major reductions in the plasma concentrations of most other amino acids (Table I), including the neutral amino acids generally believed to compete with tryptophan for uptake into the brain (Blasberg and Lajtha, 1965; Guroff and Udenfriend, 1962). Two hours after rats received the insulin, brain tryptophan levels were elevated by 36% ($P < 0.001$), and brain serotonin by 28% ($P < 0.01$) (Fernstrom and Wurtman, 1972a).

The increase in brain serotonin content observed in rats receiving insulin could have resulted not from in-

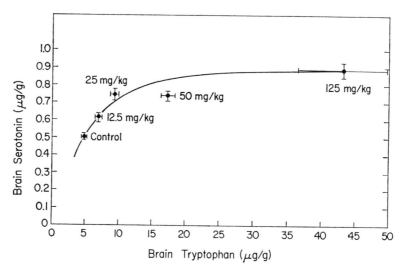

FIGURE 3 Dose-response curve relating brain tryptophan and brain 5-HT. Groups of 10 rats received tryptophan intraperitoneally at noon, and were killed 1 hr later. All brain tryptophan levels were significantly higher than control levels ($P < 0.01$). (Reproduced from Fernstrom and Wurtman, 1971b.)

TABLE I

Effect of insulin on concentrations of tryptophan, other amino acids, and glucose in rat plasma

	Concentration		
	Control	Insulin	Percent
Amino Acid	(μg/ml)		Change
L-Tryptophan	11.1 ± 0.6	16.6 ± 0.9	+50*
L-Tyrosine	11.4 ± 0.8	9.4 ± 1.0	−18
L-Phenylalanine	11.8 ± 0.3	10.8 ± 0.9	−9
L-Serine	24.6 ± 1.2	17.0 ± 1.3	−31*
L-Glycine	29.6 ± 1.4	18.3 ± 2.0	−38*
L-Alanine	25.3 ± 0.9	11.7 ± 0.5	−54*
L-Valine	21.5 ± 1.3	17.3 ± 1.8	−20
L-Isoleucine	13.7 ± 0.8	7.1 ± 0.4	−48*
L-Leucine	21.3 ± 1.2	16.6 ± 2.1	−22
	(mg/100 ml)		
Glucose	96.0 ± 2.5	43.0 ± 4.8	−55*

*$P < 0.01$.

Groups of five to ten fasting 150 to 200-g rats were killed 2 hr after receiving insulin (2 units/kg i.p.). (Reproduced from Fernstrom and Wurtman, 1972a.)

creased availability of substrate but from reflexes activated by the accompanying hypoglycemia. To determine whether the physiological secretion of insulin, in normoglycemic animals, also increases plasma and brain tryptophan concentrations and brain serotonin, these indoles were measured in rats fasted for 15 hr and then given access to a carbohydrate diet. In a typical experiment, the animals ate an average of 5 g/hr during the first hour, and 2 g/hr during the second and third hours (Fernstrom and Wurtman, 1971b, 1972a). Plasma tryptophan levels were significantly elevated 1, 2, and 3 hr after food presentation; tyrosine concentrations were depressed at all three times studied (Table II). Brain tryptophan concentrations rose 22% during the first hour, reached a peak 65% above control values ($P < 0.001$) after 2 hr, and remained significantly elevated at 3 hr (Table II). Brain serotonin concentrations rose dur-

ing the first hour, became significantly elevated by the end of the second hour, and remained so after 3 hr (Table II) (Fernstrom and Wurtman, 1971b).

On the basis of these observations, a model was constructed to explain the mechanisms by which dietary inputs affect brain serotonin. According to this model, carbohydrate consumption will, by eliciting insulin secretion, raise plasma tryptophan levels (by mobilizing the amino acid from yet undefined pools); this elevation, in turn, will cause a corresponding increase in brain tryptophan, which will increase the saturation of tryptophan hydroxylase, increase serotonin synthesis, and, ultimately, increase brain serotonin levels. On the basis of this model, we predicted that the consumption of diet containing both carbohydrates and protein would cause an even greater rise in brain serotonin. In addition to elevating plasma tryptophan via insulin secretion, the tryptophan molecules in the dietary proteins would also contribute directly to plasma tryptophan; hence plasma tryptophan concentrations would increase even more than after ingestion of a protein-free diet, and brain tryptophan and serotonin would show similar amplifications in response. When this model was tested by giving fasted rats access to diets containing either casein or a synthetic amino acid similar in composition to 18% casein, it was immediately apparent that it was in need of major revision: As expected, protein consumption was followed by a major increase (about 60%, $P < 0.001$) in plasma tryptophan; however neither brain tryptophan nor brain serotonin was at all increased (Fernstrom and Wurtman, 1972b).

Other investigators, using brain slices (Blasberg and Lajtha, 1965) or animals treated with pharmacological doses of individual amino acids (Guroff and Udenfriend, 1962) had shown that groups of amino acids (e.g., neutral, acidic, basic) are transported into brain by specific carrier systems, and that within a group, the member amino acids compete with each other for common transport

TABLE II

Effect of carbohydrate ingestion on brain 5-HT concentrations and on plasma and brain tryptophan

	Time After presentation of Food (hr)			
	0	1	2	3
Plasma tryptophan	10.86 ± 0.55	13.56 ± 0.81**	14.51 ± 0.70***	13.22 ± 0.65**
Brain tryptophan	6.78 ± 0.40	8.32 ± 0.63*	11.24 ± 0.52***	9.81 ± 0.50***
Brain 5-HT	0.549 ± 0.015	0.652 ± 0.046	0.652 ± 0.012***	0.645 ± 0.017***
Plasma tyrosine	13.03 ± 0.29	9.55 ± 0.34***	8.67 ± 0.26***	9.03 ± 0.21***

*$P < 0.05$ differs from 0-time group.
**$P < 0.02$ differs from 0-time group.
***$P < 0.001$ differs from 0-time group.

Plasma amino acid concentrations are in μg/ml. Brain tryptophan and 5-HT concentrations are in μg/g brain, wet weight. Average animal weight was 160 g. (Reproduced from Fernstrom and Wurtman, 1971b.)

sites. Since protein ingestion introduces variable amounts of all of the amino acids into the blood, it seemed possible that brain tryptophan failed to increase after protein ingestion because the plasma concentrations of other, competing, amino acids increased even more than that of tryptophan. To test this hypothesis, we allowed groups of animals to eat either a synthetic diet containing carbohydrates plus all of the amino acids in the same proportions as present in an 18% casein diet, or this diet minus five of the amino acids thought to share a common transport system with tryptophan (i.e., tyrosine, phenylalanine, leucine, isoleucine, and valine). Both diets significantly increased plasma tryptophan levels above those found in fasted controls. However, only when the competing neutral amino acids were deleted from the diet did large increases occur in brain tryptophan, serotonin, or 5-hydroxyindoleacetic acid (Figure 4).

To rule out the possibility that the increase in brain 5-hydroxyindoles observed in rats consuming the latter diet was simply a nonspecific consequence of the omission of any group of amino acids from the diet, we repeated the above experiment omitting aspartate and glutamate instead of the five neutral amino acids. [These two amino acids comprise approximately the same percent of the total alpha-amino nitrogen in casein as the five competing amino acids. Because they are charged at physiologic pH, they are transported into the brain by a carrier system different from that transporting tryptophan (Blasberg and Lajtha, 1965). Hence, their absence would not be expected to alter the postprandial competition between tryptophan and other amino acids within its transport group for uptake into the brain.] At 1 and 2 hr after presentation of this diet or the complete amino acid mixture, plasma tryptophan concentrations again increased 70 to 80% above those of fasted controls ($P < 0.001$). However, neither diet caused increases in brain tryptophan, serotonin, or 5-hydroxyindoleacetic acid.

These results were interpreted as showing that brain tryptophan and 5-hydroxyindole levels do not simply reflect plasma tryptophan but also depend upon the plasma concentrations of other neutral amino acids.

FIGURE 4   Effect of ingestion of various amino acid-containing diets on plasma and brain tryptophan, and brain 5-hydroxyindole levels. Groups of 8 rats were killed 1 or 2 hr after diet presentation. Vertical bars represent standard errors of the mean. Fasting controls: ⚲ ; complete amino acid mix diet: ▮ ; mix diet minus tyrosine, phenylalanine, leucine, isoleucine, and valine: ⬤ . 1 and 2-hr plasma tryptophan levels were significantly greater in animals consuming both diets ($P < 0.001$) than in fasting controls. All brain tryptophan, serotonin, and 5-hydroxyindoleacetic acid levels were significantly greater in rats consuming the diet lacking the 5 amino acids than in fasting controls ($P < 0.001$ for all but 1-hr serotonin, $P < 0.01$). (Reproduced from Fernstrom and Wurtman, 1972b.)

This relationship is perhaps best illustrated by a correlation analysis comparing the brain tryptophan level and the ratio of plasma tryptophan to the five competing amino acids among individual rats given various diets that contain differing amounts of each amino acid. This analysis yielded a correlation coefficient of 0.95 ($P < 0.001$ that $r = 0$), whereas the correlation between brain tryptophan

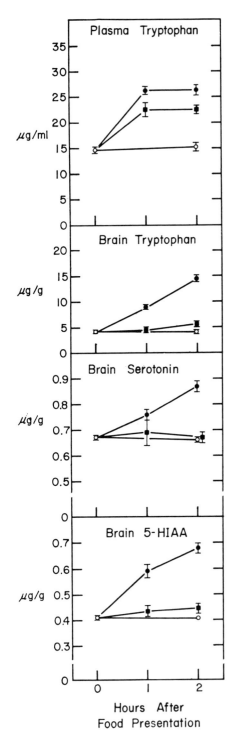

and plasma tryptophan alone was less striking ($r = 0.66$; $P < 0.001$ that $r = 0$). Similarly, the correlation coefficient for brain 5-hydroxyindoles (serotonin plus 5-hydroxyindoleacetic acid) versus the plasma amino acid ratio was 0.89 ($P < 0.001$), whereas that of 5-hydroxyindoles versus tryptophan alone was only 0.58 ($P < 0.001$) (Figure 5). Thus, the brain concentrations of both tryptophan and the 5-hydroxyindoles more nearly reflect the ratio of plasma tryptophan to competing amino acids than the plasma tryptophan concentration alone. The reason that brain tryptophan and serotonin appeared, in our earlier formulation, to depend upon plasma tryptophan alone was that all of the physiological manipulations tested at that time (i.e., tryptophan injections, insulin injections, carbohydrate consumption) raised the numerator in the plasma tryptophan:competitor ratio while either lowering the denominator or leaving it unaltered. Only when the rats consumed protein were both the numerator and the denominator elevated.

The effect of food consumption on 5-hydroxyindoles in rat brain may now be modeled as in Figure 6. Since carbohydrate ingestion elicits insulin secretion, it simultaneously raises plasma tryptophan and lowers the concentrations of the competing neutral amino acids in rats (Fernstrom and Wurtman, 1972b); hence the ratio of plasma tryptophan to competing amino acids increases, leading to elevations in brain tryptophan and serotonin. In contrast, protein consumption provides the plasma with an exogenous source of all of the amino acids; however, the ratio of tryptophan to its competitor amino acids is almost always lower in dietary proteins than it is in plasma. Probably for this reason, protein ingestion increases the plasma levels of tryptophan less than it does the concentrations of competing amino acids, and thereby

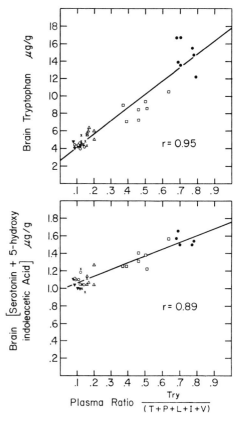

FIGURE 5 *Above:* Correlation between brain tryptophan concentration and the plasma ratio of tryptophan to the 5 competing amino acids in individual rats studied in the experiment described in Figure 3. $r = 0.95$ ($P < 0.001$ that $r = 0$). *Below:* Correlation between the sum of brain serotonin and 5-hydroxyindoleacetic acid, and the plasma ratio of tryptophan to the 5 competitor amino acids, in individual rats studied in the experiment described in Figure 1. $r = 0.89$ ($P < 0.001$ that $r = 0$). One hour control, ○; 2-hr control, ▼; 1-hr complete amino acid mix diet, ×; 2-hr complete amino acid mix diet, △; 1-hr complete mix diet minus five competing amino acids, □; 2-hr complete mix diet minus five competing amino acids, ●. (Reproduced from Fernstrom and Wurtman, 1972b.)

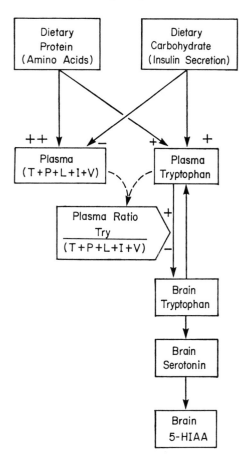

FIGURE 6 Proposed sequence describing diet-induced changes in brain serotonin concentration in the rat (see text). The ratio of tryptophan to (tyrosine + phenylalanine + leucine + isoleucine + valine) in the plasma is thought to control the tryptophan level in the brain. (Reproduced from Fernstrom and Wurtman, 1972b.)

decreases the tryptophan:competitor ratio. The insulin secretion elicited by protein consumption will, by itself, produce an opposite change in this ratio. Thus, brain tryptophan and 5-hydroxyindole levels can decrease, increase, or remain unchanged after eating, depending on the proportion of protein to carbohydrates in the diet and the amino acid composition of the particular proteins. Our most recent observations suggest that diets containing high concentrations of protein (i.e., 40% casein, or as high a proportion of protein as would be consumed in a steak) actually *decrease* brain serotonin synthesis in rats.

The extent to which these observations on rats also apply to humans and other mammals awaits clarification. Insulin does not seem to raise total plasma tryptophan in humans as it does in rats; however it does profoundly depress the concentrations of the other neutral amino acids and thus elevates the ratio of tryptophan to its competitors. In both species, the more relevant index of tryptophan availability may be plasma *free* tryptophan (i.e., tryptophan not bound to albumin). Plasma free tryptophan concentrations do fall precipitously in human subjects who consume a glucose load. When a method becomes available for estimating brain serotonin synthesis in humans, it will be interesting to determine whether such synthesis correlates best with free or total plasma tryptophan concentrations, or with their ratios to the concentrations of other neutral amino acids.

We are accustomed to believing that brain biochemistry controls behavior. The data described above show that the reverse is also true; the decision on the part of the animal to eat, and its choice of what it eats, cause characteristic changes in brain composition. This is perhaps mildly shocking, in view of the fact that the brain chemical influenced by elective food consumption (serotonin) is a neurotransmitter liberated in neuronal circuits known to control behavior. The ability of brain serotonin levels to vary in response to diet-induced changes in plasma amino acid concentrations can be viewed as a potentially dangerous curiosity that somehow managed to survive the evolutionary process. One wonders why something as important as brain neurotransmitter levels should be subject to the vagaries of what one chooses to eat for breakfast. Alternatively, this control mechanism can be viewed as having survived because it was somehow useful to the organism. We favor the latter view; we suspect that serotonin-releasing brain neurons function as a kind of humoral-neural transducer, which converts information about peripheral metabolism (i.e., as manifested by plasma amino acid levels) to neural signals (i.e., the release of a greater or lesser quantity of serotonin). Current studies are designed to identify the uses to which the brain might put this information. Perhaps the brain decides in part when it is hungry, or when it is time to sleep, to trigger ovulation, or to raise or lower the body temperature on the basis of information that it receives about plasma amino acid levels (which, in turn, depend upon food consumption). All of these brain functions have been shown by lesion or drug studies to involve serotonergic neurons. Perhaps nutritional inputs control the brain in order that the brain can make appropriate decisions concerning nutritional strategies.

Summary

There is abundant evidence that *chronic malnutrition* can alter the chemical composition of the brain: For example, inadequate protein consumption early in life is associated with long-term decreases in brain DNA and RNA contents, myelin synthesis, and catecholamine concentrations. More recent studies have shown that brain composition also responds *acutely* to *normal* nutritional inputs: The consumption of carbohydrates by the rat initiates a sequence of events that causes within an hour an increase in the brain concentration of the neurotransmitter serotonin. These events include the secretion of insulin, the resulting increases in plasma tryptophan (and decreases in plasma levels of other neutral amino acids that compete with tryptophan for uptake into the brain), and brain tryptophan, the increased saturation of the enzyme tryptophan hydroxylase, and enhanced serotonin biosynthesis. Protein consumption also raises plasma tryptophan; however, the plasma concentrations of other neutral amino acids rise even more, and thus brain tryptophan and serotonin levels are not increased. These observations indicate that *behavior* (i.e., the elective consumption of specific foods) can influence *brain biochemistry*, as well as vice versa. They also show that the nutritional status of the brain (e.g., the amounts of tryptophan and other substances available to it from the blood stream) need not parallel the general nutritional state of the body. The capacity of serotonin-containing neurons to "sense" plasma amino acid levels (i.e., by increasing their neurotransmitter content and, perhaps, secretion) suggests that they may function to monitor peripheral metabolic states, and to provide the rest of the brain with information to be used in formulating appropriate behavioral and neuroendocrine strategies.

ACKNOWLEDGMENTS These studies were supported in part by grants from the John A. Hartford Foundation and the National Aeronautics and Space Administration.

REFERENCES

ALBRECHT, P., M. B. VISSCHER, J. J. BITTNER, and F. HALBERG, 1956. Daily changes in 5-hydroxytryptamine concentration in mouse brain. *Proc. Soc. Exp. Biol. Med.* 92:702–706.

BLASBERG, R., and A. LAJTHA, 1965. Substrate specificity of steady-state amino acid transport in mouse brain slices. *Arch. Biochem. Biophys.* 112:361–377.

FERNSTROM, J. D., and R. J. WURTMAN, 1971a. Brain serotonin content: Dependence on plasma tryptophan levels. *Science* 173:149–152.

FERNSTROM, J. D., and R. J. WURTMAN, 1971b. Brain serotonin content: Increase following ingestion of carbohydrate diet. *Science* 174:1023–1025.

FERNSTROM, J. D., and R. J. WURTMAN, 1972a. Elevation of plasma tryptophan by insulin in the rat. *Metabolism* 21:337–343.

FERNSTROM, J. D., and R. J. WURTMAN, 1972b. Brain serotonin content: Physiological regulation by plasma neutral amino acids. *Science* 178:414–416.

FISHMAN, B., R. J. WURTMAN, and H. N. MUNRO, 1969. Daily rhythms in hepatic polysome profiles and tyrosine transaminase activity: Role of dietary protein. *Proc. Nat. Acad. Sci. USA* 64:677–682.

GUROFF, G., and S. UDENFRIEND, 1962. Studies on aromatic amino acid uptake by rat brain in vivo. *J. Biol. Chem.* 237:803–806.

LOVENBERG, W., E. JEQUIER, and A. SJOERDSMA, 1968. A tryptophan hydroxylation in mammalian systems. *Adv. Pharmacol.* 6A:21–36.

MARLISS, E. B., T. T. AOKI, R. H. UNGER, J. S. SOELDNER, and G. F. CAHILL, 1970. Glucagon levels and metabolic effects in fasting man. *J. Clin. Invest.* 49:2256–2270.

MOIR, A. T. B., and D. ECCLESTON, 1968. The effects of precursor loading in the cerebral metabolism of 5-hydroxyindoles. *J. Neurochem.* 15:1093–1107.

WOOL, I. G., 1965. Relation of effects of insulin on amino acid transport and on protein synthesis. *Fed. Proc.* 24:1060–1070.

WURTMAN, R. J., 1970. Diurnal rhythms in mammalian protein metabolism. In *Mammalian Protein Metabolism*, H. N. Munro, ed. New York: Academic Press, Vol. IV, Chap. 36.

WURTMAN, R. J., and J. D. FERNSTROM, 1972. L-tryptophan, L-tyrosine, and the control of brain monoamine biosynthesis. In *Perspectives in Neuropharmacology*, S. H. Snyder, ed. Oxford: Oxford University Press, pp. 145–193.

WURTMAN, R. J., C. M. ROSE, C. CHOU, and F. LARIN, 1968. Daily rhythms in the concentrations of various amino acids in human plasma. *New Eng. J. Med.* 279:171–175.

Brain Dopamine Neurons and Behavior

URBAN UNGERSTEDT

ABSTRACT The importance of nigrostriatal dopamine neurons in the control of behavior is evaluated by inducing hypofunction (6-hydroxydopamine induced lesions) or hyperfunction (dopamine receptor stimulation) in the system. Hypofunction induces a severe state of adipsia and aphagia that seems to be secondary to a pronounced sensory neglect where the animal is unable to respond to external (and internal?) stimuli. Severe unilateral hypofunction due to a 6-hydroxydopamine lesion blocks ipsilateral but not contralateral electrical self stimulation. Hyperfunction due to pharmacologically induced receptor stimulation brings the animal through a series of behavioral patterns that all seem to be related to a basic behavioral repertoire. The behavioral syndromes elicited by hypofunction and hyperfunction are partly quantified in tests designed to reveal changes in dopamine function.

AMONG THE transmitter, or transmitter candidates, in the central nervous system, dopamine (DA) is uniquely related to known behavioral syndromes. We know that a degeneration of the nigrostriatal DA system is the major functional brain lesion underlying Parkinson's disease (Hornykiewicz, 1966), and the disease is therefore successfully treated by substituting for the lost DA with high doses of its precursor DOPA (Cotzias et al., 1967). Important parts of the Parkinson symptomatology may be reproduced in animals after lesions of the DA pathway (Battista et al., 1969; Sourkes et al., 1969). These animal models are valuable tools in our attempts to understand the nature of the functional lesion.

Biochemistry and histochemistry form the basis for our understanding of the part played by DA neurons in brain function, i.e., the biochemical identification of DA (Carlsson et al., 1958) and the histochemical identification of the DA neurons (Carlsson et al., 1962; Andén et al., 1965). Biochemical and anatomical findings have been linked to behavior mainly through DA pharmacology; e.g. the depletion of DA stores is probably the major reason for the pronounced behavioral depression caused by reserpine.

Conversely, the behavioral excitation and stereotyped behavior seen after amphetamine may be due to excessive release of DA (Randrup and Munkvad, 1966). The psychotic syndrome caused by an overdose of amphetamine in man looks very similar to acute schizophrenia (Bell, 1965; Snyder, 1972) and to complete the parallel, neuroleptic drugs, effective against schizophrenia, are found to inhibit DA receptors (Andén et al., 1970).

The fact that pronounced behavioral changes take place after changes in DA neurotransmission seems to provide us with a unique chance to link behavior to synaptic transmission within a defined neuron system in the brain. This article relates a series of experiments where we have tried to describe and interpret behavior in relation to experimental manipulation of DA anatomy and pharmacology.

In the rat brain, it is possible to distinguish two different ascending DA neuron systems, the nigrostriatal system and the mesolimbic system (Andén et al., 1966; Ungerstedt, 1971a) (Figure 1). The *nigrostriatal DA system* originates from cell bodies in the zona compacta of the substantia nigra. Its axons run rostromedially to enter the lateral aspect of the medial forebrain bundle. In the lateral hypothalamus, the axons run in a dense bundle that moves laterally to enter the internal capsule, fan out in the globus pallidus, and terminate in the nucleus caudatus-putamen, i.e., the striatum.

The *mesolimbic DA system* originates from cell bodies situated dorsal and lateral to the interpeduncular nucleus. The axons run rostrolaterally to enter the medial forebrain bundle. They run slightly medial to the nigrostriatal axons and never enter into the internal capsule but continue to the level of the anterior commissure, where they terminate in the nucl. accumbens and the olfactory tubercle.

In order to reveal the role played by central DA neurons in the regulation of behavior, we have used lesions of the DA pathways and pharmacological interference with their synaptic transmission. Local injections of 6-hydroxydopamine (6-OH-DA) into areas containing DA neurons have proved to be a very effective and reproducible way of performing lesions (Ungerstedt, 1968, 1971b). The lesions are far more selective in their effect on DA and norepinephrine (NE) neurons than conventional lesions (Ungerstedt, 1971b, 1973). The detailed mapping of the monoamine pathways in the rat brain (Ungerstedt, 1971a) has provided the necessary anatomical knowledge to localize injection cannulas in the DA pathways.

URBAN UNGERSTEDT Department of Histology, Karolinska Institutet, Stockholm, Sweden

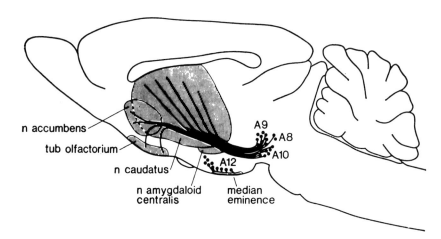

FIGURE 1   Sagittal projection of the brain dopamine pathways. The nigrostriatal system originates in the cell group labeled A9, while the mesolimbic system originates in the group labeled A10. Group A8 probably innervates the striatum while the A12 innervates the median eminence in the hypothalamus. (From Ungerstedt, 1971a.)

In the further analysis of the lesioned animals we have attempted to quantify behavior that is elicited by hypo- or hyperfunction of the DA system.

The unilateral dopamine denervation syndrome

A unilateral injection of DA into the striatum produces an asymmetry in movements and posture (Ungerstedt et al., 1969). The animal deviates and even rotates in circles away from the side getting the injection. The unilaterally increased concentration of DA tends to force the animal in the direction of the side where there is less DA. A unilateral 6-OH-DA induced degeneration of the nigrostriatal DA system induces a similar syndrome. As soon as the animal wakes up from the operation, it assumes an asymmetrical posture with tail and head deviating toward the side of the lesion. The legs on the unlesioned side are usually slightly extended while the legs on the lesioned side are kept flexed under the body. Arousal of the animal by pinching its tail causes rotation toward the lesioned side. However, starting about 24 hr after the lesion, the animal develops a tendency to turn spontaneously toward the normal side. If the animal is treated with a mono-amine oxidase inhibitor at the time of the operation, it shows rotation toward the normal side at this time point, i.e., between 24 to 34 hr after the 6-OH-DA injection (Ungerstedt, 1971d). This behavior coincides with the period when DA disappears from the degenerating neurons (Hökfelt and Ungerstedt, 1969). This is the behavioral expression of the degeneration release of DA, comparable to the degeneration contraction of the nictitating membrane after the extirpation of the superior cervical ganglion (Langer, 1966).

The DA nerve terminals have lost their transmitter content 48 hr after the 6-OH-DA lesion as determined with histochemical (Ungerstedt, 1971d) and electron-microscopic technique (Hökfelt and Ungerstedt, 1969). The animal now enters into a chronic state where the unilateral loss of DA expresses itself in several ways: The animal shows a chronic deviation towards the side of the lesion. Spontaneous exploratory behavior is decreased, and there is a transient period of adipsia and aphagia (2 to 3 days). There are also signs of upset vegetative function as the animals show a chronic loose consistency of their feces. However, the severity of the syndrome becomes apparent only after an examination of the sensory functions of the animal: A unilateral degeneration of the nigrostriatal DA system produces a severe "sensory neglect" on the side of the body contralateral to the lesion (Ljungberg and Ungerstedt, to be published). The animal fails to orient towards tactile, auditory, olfactory, and visual stimuli applied to that side of the body. The syndrome has been characterized by the orienting reaction evoked in a series of tests where different sensory stimuli have been applied to both sides of the body, e.g. a straw pressed against the body, whisker touch, a piece of cotton soaked in a smelling solution moved toward a nostril from behind, a sound evoked close to one ear and, to test vision, the placing reaction of the fore paw when the animal is held in the experimenter's hand and moved past the edge of a table, or simply the orienting reaction to an object moved from behind into the field of vision of the animal. While the animal shows essentially normal reaction to stimuli applied ipsilateral to the degeneration of the DA system, there is initially an almost total irresponsivity to stimuli applied on the side contralateral to the lesion. There is a gradual recovery of some sensory abilities.

After $1\frac{1}{2}$ weeks the rats react to smell, and after 7 weeks most of the animals react to visual stimuli, while normal reactions to tactile stimuli have not returned even 3 months after the lesion.

The fact that the animal initially shows a pronounced asymmetry in its movements, which recover only slightly, seems to suggest that the deficit in orienting reaction might be a motor rather than a sensory deficit. However, the fact that there is a recovery of some sensory modalities, e.g. smell, while others, e.g. tactile, remain deficient suggests that the animal has no motor difficulties in performing the orienting reaction per se. The lack of orienting reaction is more likely to be due to a deficiency in the ability to recognize the stimulus or to integrate sensory and motor function to perform an adequate response.

If a unilaterally DA denervated animal is given drugs that change DA transmission this will change the degree of asymmetry in posture and movements. Amphetamine, known to release DA (Glowinski and Axelrod, 1965; Carlsson et al., 1966), will increase the asymmetry to the point of vigorous rotation toward the lesioned side (Ungerstedt and Arbuthnott, 1970; Ungerstedt, 1971c). This agrees with the previously related study where DA was injected unilaterally into the striatum and produced a similar but less pronounced rotation away from the injected side (Ungerstedt et al., 1969). As this rotational behavior in all probability is quantitatively related to the degree of DA receptor stimulation, we constructed a "rotometer" where the number of full turns per minute may be recorded over long periods of time (Ungerstedt and Arbuthnott, 1970). The rotometer consists of a plexiglass bowl shaped as a hemisphere in which the animal moves around connected by a thin wire to a microswitch arrangement localized in the geometrical center of the sphere. Every full turn is registered on electromechanical counters or cumulative recorders. Drugs, like amphetamine, that release DA from the presynaptic DA nerve terminals, will work only on the remaining DA system and thus induce rotation toward the side of the degenerated DA system (Ungerstedt, 1971c). Postsynaptically acting drugs, like apomorphine, that stimulate DA receptors (Ernst, 1967; Andén et al., 1967), will act on both innervated and denervated DA receptors. Apomorphine causes a rotation in the direction opposite to amphetamine. The animal rotates toward the normal side. This indicates that apomorphine preferentially stimulates the denervated DA receptors, i.e., they seem to become supersensitive after denervation (Ungerstedt, 1971d; Schoenfeld and Uretsky, 1972). The animals react to DOPA in the same way as to apomorphine. DOPA is converted to DA by the enzyme still remaining in the denervated striatum, and because of the supersensitivity the small amounts of DA thus formed may compensate for the loss of endogenous DA. Postsynaptic supersensitivity to DA may form the basis for the therapeutic effect of DOPA in Parkinson's disease (Ungerstedt, 1971e).

The bilateral dopamine denervation syndrome

Bilateral removal of the nigrostriatal DA systems by intracerebral injections of 6-OH-DA, produces a very severe syndrome, as might be expected from the findings in the unilaterally denervated animals. The animals are hypokinetic, adipsic, and aphagic immediately after the operation. In our initial studies the condition persisted for about 3 to 4 weeks except for a period of increased activity coinciding with the release of DA from the degenerating neurons (Ungerstedt, 1970, 1971f). If not tube fed, the animals die within 3 to 5 days after the operation. The syndrome is very similar, or identical, to the "lateral hypothalamic syndrome" that follows destruction of an area close to the crus cerebri in the lateral hypothalamus. This area has been described as the "lateral hypothalamic eating center" and is thought to be involved in the regulation of eating and drinking (Anand and Brobeck, 1951). However, the stereotaxic mapping of the ascending DA axons has revealed that the axons run through the lateral hypothalamus assembled in a dense bundle at the tip of the crus cerebri (Ungerstedt, 1971f) (Figure 2). A lesion of this area interrupts the DA pathway as effectively as the 6-OH-DA injection into the substantia nigra and denervates the striatum of its DA nerve terminals. Several authors have argued for the possibility that lesions of the lateral hypothalamic area interrupt axons passing through this area (Morgane, 1961; Gold, 1967; Albert et al., 1970). Apart from ourselves (Ungerstedt, 1970, 1971f) Oltmans and Harvey (1972) have suggested that the crucial damage is to the nigrostriatal DA system. Recent studies with intraventricular injection of 6-OH-DA have also argued in favor of this hypothesis (Zigmond and Stricker, 1972).

The lateral hypothalamic syndrome has often been described only as an adipsic and aphagic syndrome. However, several authors have commented on symptoms other than deficits in eating and drinking, such as hypokinesia, catalepsia, decreased exploratory behavior, and arousal difficulties (Gladfelter and Brobeck, 1962; Balagura et al., 1969; Beattie et al., 1969). This other aspect of the lateral hypothalamic syndrome is especially evident in a larger animal such as the dog (Fonberg, 1969). These findings make the parallel to what is found in the bilaterally DA denervated animals even more striking. In a recent report, Marshall et al. (1971) have described a pronounced sensory neglect after lateral hypothalamic lesions. Our own findings in the DA denervated animal (Ljungberg and

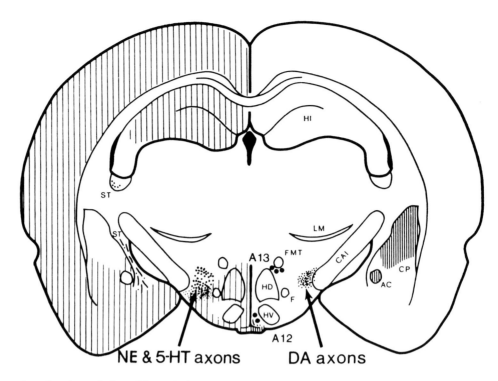

NE & 5-HT axons   **DA axons**

FIGURE 2   Frontal section through the midhypothalamus of a rat. The ascending monoamine axon bundles are indicated by dots. The dopamine axons run in a dense bundle in the lateral hypothalamus. The norepinephrine and the 5-hydroxytryptamine (fine dots) axons follow a more medial course. (From Ungerstedt, 1971f.)

Ungerstedt, to be published, see above) confirm their findings and provide additional evidence that the lateral hypothalamic syndrome may be attributed to a degeneration of the nigrostriatal DA system. The sensory neglect resulting from a degeneration of the nigrostriatal DA system probably explains the severe syndrome of adipsia and aphagia. The animal is essentially unable to respond adequately to any sensory stimuli, possibly neither internal (e.g. proprioceptive) nor external. The adipsia and aphagia syndrome that has a recovery period of a few weeks is not necessarily associated with catalepsia, i.e., the inability of the animal to escape from posture it has been passively positioned in by the experimenter. This was concluded by Balagura et al. (1969) after studies of lateral hypothalamic lesion and by ourselves after lesions of the DA system (Ungerstedt, 1971f). However, in a recent study (Avemo et al., to be published), we have attempted to produce as extensive a DA degeneration as possible, i.e., we have disposed our 6-OH-DA into the bundle of DA axons where it leaves the substantia nigra rather than into the substantia nigra itself. This allows for a more complete degeneration of all DA fibers. We have tube fed such animals for 3 to 4 months without any sign of recovery. Such lesions show histochemically a complete degeneration of both the limbic and the striatal DA system. The animals are catalepsic, akinetic, rigid, and

show a fine tremor of high frequency. The motor phenomena are obviously well related to the Parkinson syndrome in man. However, there is still some uncertainty whether this severe syndrome is solely due to a lesion of DA neurons, or if the small unspecific lesion that always occurs after 6-OH-DA injections (Ungerstedt 1971b,1973) damages other neurons that contribute to the behavioral syndrome.

In order to further test the bilateral DA denervation syndrome, we have developed certain behavioral tests related to its typical symptomatology: The animals are tested in an automatic "hole-board" (Ljungberg and Ungerstedt, to be published) which consists of a 70 × 70 cm square, open field. Eighty-one 3.5 cm wide, round holes are distributed evenly in the floor. The open field activity is monitored by photocell beams 3 cm above the floor. Hole exploration is monitored by photocells that react every time the animal peeps into a hole (Figure 3).

A normal animal explores the hole-board for about 15 min by looking into holes and walking around. The bilaterally DA denervated animal shows a complete lack of exploratory behavior. Apomorphine, however, induces a strong activity in the animal even at a very low dose (0.05 mg/kg). This is the same phenomenon as the rotation of the unilaterally denervated animal and is explained by the DA degeneration supersensitivity to apomorphine (Ungerstedt,1971c; Schoenfeld and Uretsky,

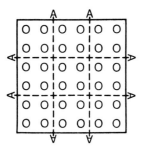

FIGURE 3 Schematic drawing of the "hole-board" seen from above. Photocells cover the open field area as well as each hole (not shown).

1972). The bilaterally denervated animal shows no reaction to amphetamine (2 mg/kg of *d*-amphetamine) (Figure 4), which also corresponds to the finding in the rotometer where the direction of the amphetamine induced rotation indicates that the amphetamine effect is elicited only from the intact DA neurons.

There is an interesting correlation between the severity of the adipsia and aphagia and the response to amphetamine and apomorphine (Avemo et al., to be published). Animals showing a severe and long-lasting adipsia and aphagia do not react to amphetamine but show a strongly increased reaction to apomorphine, while animals that show a fast recovery after the adipsia and aphagia react to amphetamine but less to apomorphine. Brought together with our previous results on adipsia and aphagia (see above), it seems as if the amphetamine induced excitation occurs only in the presence of remaining DA neurons.

However, only a fraction of the normal DA innervation seems necessary to elicit an amphetamine response or to allow for a recovery from the adipsia and aphagia.

The role of the NE neurons remains somewhat unclear in the adipsic and aphagic syndrome as well as in the amphetamine induced excitation. The 6-OH-DA induced degeneration of the DA neurons causes a partial denervation of forebrain NE nerve terminals. As involvement of NE neurons has been suggested in both the adipsic and aphagic syndrome and the amphetamine induced excitation, we have attempted to control for the damage of ascending NE pathways by injecting 6-OH-DA directly into these pathways. This causes degeneration of forebrain NE neurons without affecting the DA neurons. Such animals show no sign of adipsia and aphagia (Ungerstedt, 1971); in fact there are even signs of overeating. In the hole-board there is a slight modification of the ratio between walking around and looking into holes but no sign of a decreased amphetamine response. These results make it possible to exclude a damage of NE neurons as an important factor in the behavioral changes occurring after 6-OH-DA lesions of the DA neurons.

There are several pharmacological studies indicating differences in the way NE and DA neurons influence behavior (see Randrup and Munkvad, 1970). The NE neurons are thought to influence locomotor behavior, while the DA neurons are thought to influence stereotyped behavior. However, our own studies in the hole-board indicate that this may be too simple a way to look upon the various components of behavior. Amphetamine

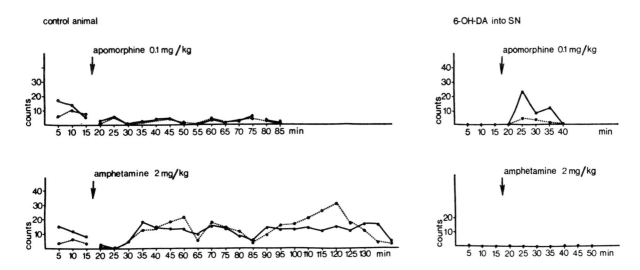

FIGURE 4 Recordings from the hole-board. The control animal shows normal response to amphetamine and no response to the subthreshold dose of apomorphine. The animal with a bilateral lesion of the nigrostriatal dopamine system does not respond to amphetamine while it shows a clear cut reaction to the low dose of apomorphine. Solid lines show the movements in the open field while dotted lines show the peeping into the holes.

(2 mg/kg *d*-amphetamine), which releases both NE and DA (Carlsson et al., 1966), causes varied behavior where the animals walk around and look into the holes frequently. The behavior is similar to the initial behavior shown by a normal animal that explores the hole-board for the first time. However, apomorphine (1 to 2 mg/kg), which stimulates DA receptors (Ernst, 1967; Andén et al., 1967), induced a different pattern. Looking into holes is completely abolished while the walking around is greatly increased (Figure 4). In fact, there is a gradual change in the behavior when the dose of apomorphine is increased (Ljungberg and Ungerstedt, to be published). Low doses (0.1 to 0.5 mg/kg) inhibit normal behavior and cause "freezing." In doses varying between 1 and 2 mg/kg, there is a great increase in locomotor behavior, while, at higher doses, locomotor behavior decreases and oral behaviors, such as licking and gnawing, dominate. There is a great increase in sniffing except during the stage of freezing. The higher doses of apomorphine (5 mg/kg) often elicit aggressive behavior.

It is fascinating to note that the increasing doses of apomorphine seem to elicit different patterns of behavior that when brought together in a meaningful order seem to be components of a "survival behavior," i.e., locomotion, sniffing, licking, and gnawing are all components of food seeking and ingestion, and together with freezing and aggression they seem to incorporate many of the basic components of normal behavior.

The bilaterally DA denervated animals do not show any spontaneous activity. In order to study the performance of such a severely hypokinetic "nonperforming" animal, we have developed an under-water Y-maze (Ranje and Ungerstedt, to be published). The animals are brought under water in a start box, and swim to the choice point of the Y where they learn to select either a lighted or a darkened arm in order to reach the water surface so as to breathe. The swimming time is measured and the number of mistakes counted (defined as entries into the wrong arms of the maze). In this swim-maze, the lesioned animals are forced to act or they will not survive. In fact, the DA denervated animals are able to swim the maze and to learn to select the lighted or the darkened arm in order to reach the surface. As a parallel to the DA denervated animals, we have tested animals treated with DA receptor blocking drugs. Even if such animals show the most serious catalepsia, they swim the maze just as a normal animal. It is evident that the severe behavioral inhibition induced by blockade of the DA transmission may be overcome during certain types of stress. The phenomenon seems to be a parallel to the "paradoxical kinesia" shown even by severely disabled Parkinson patients during similarly stressful situations.

Dopamine neurons and self-stimulation

High rates of self-stimulation may be elicited from areas known to contain DA neurons, e.g. the area around the interpeduncular nucleus and the lateral hypothalamus. However, there also seems to be good evidence for the participation of NE neurons in self-stimulation (Stein, 1962; Poschel and Niuteman, 1963). Crow (1972) suggested that there, in fact, are two systems underlying self-stimulation, the NE and the DA system. However, Breese et al. (1971) and Stein and Wise (1971) managed to abolish self-stimulation by interventricular injection of 6-OH-DA and attributed these effects to a degeneration of the NE neurons. Antelman et al. (1972) reinterpreted Stein's data and showed that self-stimulation may still be elicited by priming the animals with free stimulation, which will make them resume self-stimulation. Since Breese et al. (1971) obtained the same decrease of both NE and DA after treatment with 6-OH-DA, it seems to be difficult to attribute the block of self-stimulation to one or the other of the two amines.

The major problem in using lesions to abolish a functional effect is obviously the possible compensatory postsynaptic supersensitivity (Ungerstedt, 1971d) that may help to keep up function in spite of a severe reduction in the number of presynaptic nerve terminals. In order to determine the importance of DA neurons in self-stimulation behavior, we have used a criterion to select the most completely denervated animals and then test them for self-stimulation. A large number of unilaterally DA denervated animals were tested in the rotometer on apomorphine. The rotational behavior that results reveals the extent of DA denervation (Ungerstedt, 1971c). Intense rotation indicates strong postsynaptic supersensitivity that in turn indicates extensive degeneration of the presynaptic DA nerve terminals. Fifteen animals that showed intense rotational behavior were implanted with bilateral hypothalamic electrodes (Christie et al., 1973). All 15 showed self-stimulation from the electrode situated on the normal side of the brain, while 4 failed to self-stimulate from the electrode situated on the side where the DA neurons were lesioned. These 4 animals had shown the strongest rotational behavior to apomorphine among the 15 implanted animals. The histochemistry revealed a partial denervation of ascending NE fibers, while there was a complete, or next to complete, denervation of the DA fibers. The most extensive denervation of DA was associated with the strongest rotational response to apomorphine. Those animals that showed the most extensive DA degeneration, as revealed by the rotometer test for supersensitivity and histochemistry, did not self-stimulate. This study shows a direct correlation between the ability to self-stimulate and the presence of DA nerves. The results strongly suggest that the presence of DA neurons are necessary for self-stimulation, although they don't rule out a participation of other types of neurons, e.g. NE neurons.

DA neurons and the function of the basal ganglia

The function of the central DA pathways have to be understood in connection with the function of the areas they innervate. The nigrostriatal pathway innervates the nucl. caudatus putamen (striatum). The mesolimbic DA system innervates the olfactory tubercle, the nucl. accumbens and parts of the septal area and may be regarded as part of the limbic system. So far there are few behavioral changes that can be attributed solely to the limbic DA system (see Fuxe and Ungerstedt, 1970), and in the following we are forced to consider mainly the nigrostriatal system and the basal ganglia.

There is no single, known neuron system in the brain that after experimental manipulation may influence behavior as profoundly as the nigrostriatal DA system. The behavioral syndromes range from extreme hyper-activity to severe akinesia and sensory neglect. However, there seem to be several mechanisms aimed at main-taining "homeostasis" within the system. A pharma-cological increase in DA receptor excitation by apomor-phine is counteracted by a decreased turnover of the endogenous transmitter and, conversely, a pharmaco-logical decrease of the transmission by a DA receptor blocking drug evokes an increase in the turnover of DA (Carlsson and Lindqvist, 1963; Andén et al., 1967). In the case of neuronal degeneration of the DA system, function is maintained after severe but incomplete loss of nerve terminals by a compensatory postsynaptic super-sensitivity (Ungerstedt, 1971d).

From the anatomy of the nigrostriatal DA system (see Ungerstedt, 1971a), we know that changes in DA trans-mission must affect behavior by changing the activity of the striatum. Most available evidence indicates that DA is inhibitory in the striatum: Iontophoretically applied DA inhibits striatal cells (Bloom et al., 1965), and electrical stimulation of the substantia nigra exerts an inhibitory reaction in the same way as iontophoretically applied DA (Connor, 1970). Recent studies by Feltz and de Champlain (1972) show that the inhibitory effect of stimulating the substantia nigra may be abolished by 6-OH-DA treatment, which probably lesions the DA neurons. As might be expected from these studies, de-generation of the nigrostriatal DA system causes an increased spontaneous firing frequency of the striatal cells (Hoffer et al., to be published). We may, thus, assume that the behavioral excitation that follows increased release of DA is due to an *inhibition* of the striatum, while the behavioral depression following a decrease in DA transmission is due to an excitation of the striatum. In agreement with this, electrical excitation of the striatum via an implanted electrode, causes a behavioral "arrest reaction" (Akert and Anderson, 1951; see also Laursen, 1963), which may possibly be the same as a short lasting behavioral depression.

The mechanisms by which the striatum influences behavior are far from understood. Anatomically, the striatum is strategically positioned to receive information from the entire cerebral cortex through the massive corticostriate fibers. The thalamus is another major source of striatal afferents, and the nigrostriatal DA system forms a third important group of afferent fibers. The major striatal efferents terminate in the substantia nigra and the globus pallidus. Pallidal efferents end in the mid-brain and in nuclei in the thalamus, which in turn are closely related to the motor cortex (see Nauta, 1969). These anatomical data seem to be in agreement with electrophysiological data indicating a "striato-thalamo-cortical" loop (Buchwald et al., 1961), possibly exerting control over sensory and motor phenomena. The be-havioral deficits following lesions of the striatum have been interpreted as due to damage to a sensory-motor integrating mechanism (see Laursen, 1963; Gybels et al., 1967). In fact, the multisensory convergence of visual, auditory, and somatic stimuli on single caudate cells supports the idea that the striatum has important inte-grative properties (see Krauthamer and Albe-Fessard, 1965).

Lesions or pharmacological manipulations of the nigro-striatal DA system may, thus, interfere with important mechanisms of sensory-motor integration by changing the activity of the striatum. The whole range of be-havioral syndromes, e.g. the rotational behavior, the sensory neglect, the various levels of hypo- and hyper-activity, and the interruption of self-stimulation behavior, feeding and drinking, may, thus, be signs of changes in the ability to perceive or respond to sensory stimuli.

ACKNOWLEDGMENT This investigation was supported by a grant from the Swedish Medical Research Council (04X-357Y).

REFERENCES

AKERT, K., and B. ANDERSON, 1951. Experimenteller Beitrag zur Physiologie des Nucleus Caudatus. *Acta Physiol. Scand.* 22:281–298.

ALBERT, D. J., L. H. STORLIEN, D. J. WOOD, and G. K. EHMAN, 1970. Further evidence for a complex system controlling feeding behavior. *Physiol. Behav.* 5:1075–1082.

ANAND, B. K., and J. R. BROBECK, 1951. Hypothalamic control of food intake in rats and cats. *Yale J. Biol. Med.* 24:123–140.

ANDÉN, N.-E., S. G. BUTCHER, H. CORRODI, K. FUXE, and U. UNGERSTEDT, 1970. Receptor activity and turnover of dopamine and norepinephrine after neuroleptics. *Europ. J. Pharmacol.* 11:303–314.

ANDÉN, N.-E., A. CARLSSON, A. DAHLSTRÖM, K. FUXE, N.-Å HILLARP, and K. LARSSON, 1964. Demonstration and mapping out of nigro-neostriatal dopamine neurons. *Life Sci.* 3:523–530.

ANDÉN, N.-E., A. DAHLSTRÖM, K. FUXE, and K. LARSSON, 1965. Further evidence for the presence of nigro-neostriatal dopamine neurons in the rat. *Amer. J. Anat.* 116:329–333.

ANDÉN, N.-E., A. DAHLSTRÖM, K. FUXE, L. LARSSON, L. OLSON, and U. UNGERSTEDT, 1966. Ascending monoamine neurons to the telencephalon and diencephalon. *Acta Physiol. Scand.* 67: 313–326.

ANDÉN, N.-E., A. RUBENSON, K. FUXE, and T. HÖKFELT, 1967. Evidence for dopamine receptor stimulation by apomorphine. *J. Pharm. Pharmacol.* 19:627–629.

ANTELMAN, S. M., A. S. LIPPA, and A. E. FISHER, 1972. 6-Hydroxydopamine, noradrenergic reward and schizophrenia. *Science* 175:919–920.

BALAGURA, S., R. H. WILCOX, and D. V. COSCINA, 1969. The effect of diencephalic lesions on food intake and motor activity. *Physiol. Behav.* 4:629–633.

BATTISTA, A. F., M. GOLDSTEIN, S. NAKATANI, and B. ANAGNOSTE, 1969. The effects of ventrolateral thalamic lesions on tremor and the biosynthesis of dopamine in monkeys with lesions in the ventromedial tegmentum. *J. Neurosurg.* 31: 164–171.

BEATTIE, C. W., H. I. CHERNOW, P. S. BERNARD, and F. H. GLENNY, 1969. Pharmacological alteration of hyperreactivity in rats with septal and hypothalamic lesions. *Int. J. Neuropharmacol.* 8:365–371.

BELL, D. S., 1965. Comparison of amphetamine psychosis and schizophrenia. *Brit. J. Psychiat.* 111:701–707.

BLOOM, F. E., E. COSTA, and G. C. SALMOIRAGHI, 1965. Anesthesia and the responsiveness of individual neurons of the caudate nucleus of the cat to acetylcholine, norepinephrine and dopamine administered by microelectrophoresis. *J. Pharmacol. Exp. Ther.* 150:244–252.

BREESE, G. R., J. L. HOWARD, and J. P. LEAHY, 1971. Effect of 6-hydroxydopamine on electrical self-stimulation. *Br. J. Pharmac.* 42:88–99.

BUCHWALD, N. A., E. J. WYERS, C. W. LANPRECHT, and G. HEUSER, 1961. The "caudate-spindle". IV. A behavioral index of caudate-induced inhibition. *Electroenceph. Clin. Neurophysiol.* 13:531–537.

CARLSSON, A., B. FALCK, and N.-Å. HILLARP, 1962. Cellular localization of brain monoamines. *Acta Physiol. Scand.* (Suppl. 196) 56:1–28.

CARLSSON, A., K. FUXE, B. HAMBERGER, and M. LINDQVIST, 1966. Biochemical and histochemical studies on the effects of imipramine-like drugs and (+)-amphetamine on central and peripheral catecholamine neurons. *Acta Physiol. Scand.* 67: 481–497.

CARLSSON, A., and M. LINDQVIST, 1963. Effect of chlorpromazine or haloperidol on formation of 3-methoxytyramine and normetanephrine in mouse brain. *Acta Pharmacol.* (*Kbh.*) 20: 140–144.

CARLSSON, A., M. LINDQVIST, T. MAGNUSSON, and B. WALDECK, 1958. On the presence of 3-hydroxytyramine in brain. *Science* 127:471.

CHRISTIE, J., T. LJUNGBERG, and U. UNGERSTEDT, 1973. Dopamine neurons and electrical self-stimulation in the lateral hypothalamus.

CONNOR, J. D., 1970. Caudate nucleus neurons: Correlation of the effects of substantia nigra stimulation with iontophoretic dopamine. *J. Physiol.* (*Lond.*) 208:691–703.

COTZIAS, G. C., M. H. VAN WOERT, and L. M. SCHIFFER, 1967. Aromatic amino-acids and modification of Parkinsonism. *New Engl. J. Med.* 276:374–379.

CROW, T. J., 1972. A map of the rat mesencephalon for electrical self-stimulation. *Brain Res.* 36:265–273.

ERNST, A. M., 1967. Mode of action of apomorphine and dexamphetamine on gnawing compulsion in rats. *Psychopharmacologia* (*Berl.*) 10:316–323.

FELTZ, P., and J. DE CHAMPLAIN, 1972. Persistence of caudate unitary responses to nigral stimulation after destruction and functional impairment of the striatal dopaminergic terminals. *Brain Res.* 43:595–600.

FONBERG, E., 1969. The role of the hypothalamus and amygdala in food intake alimentary motivation and emotional reactions. *Acta Biol. Exp.* 29:335–358.

FUXE, K., and U. UNGERSTEDT, 1970. Histochemical, biochemical and functional studies on central monoamine neurons after acute and chronic amphetamine administration. In *Amphetamines and Related Compounds*, E. Costa and S. Garattini, eds. New York: Raven Press, pp. 257–288.

GLADFELTER, W. E., and J. R. BROBECK, 1962. Decreased spontaneous locomotor activity in the rat induced by hypothalamic lesions. *Amer. J. Physiol.* 203:811–817.

GLOWINSKI, J., and J. AXELROD, 1965. Effect of drugs on the uptake release and metabolism of H^3-norepinephrine in the rat brain. *J. Pharmacol.* 149:43–49.

GOLD, R. M., 1967. Aphagia and adipsia following unilateral and bilateral asymmetrical lesions in rats. *Physiol. Behav.* 2: 211–220.

GYBELS, J., M. MEULDERS, M. CALLENS, and J. COLLE, 1967. Disturbance of visuo-motor integration in cats with small lesions of the caudate nucleus. *Arch. Internat. Physiol. Biochem.* 75:283–302.

HÖKFELT, T., and U. UNGERSTEDT, 1969. Electron and fluorescence microscopical studies on the nucleus caudatus putamen of the rat after unilateral lesions of ascending nigro-neostriatal dopamine neurons. *Acta Physiol. Scand.* 76:415–426.

HORNYKIEWICZ, O., 1966. Dopamine (3-hydroxytyramine) and brain function. *Pharmacol. Rev.* 18:925–964.

KRAUTHAMER, G., and D. ALBE-FESSARD, 1965. Inhibition of nonspecific sensory activities following striopallidal and capsular stimulation. *J. Neurophysiol.* 28:100–124.

LANGER, S. Z., 1966. The degeneration contraction of the nictitating membrane in the unanesthetized cat. *J. Pharmacol. Exp. Ther.* 151:66–72.

LAURSEN, A. M., 1963. Corpus striatum. *Acta Physiol. Scand.* 59: 1–106.

MARSHALL, J. F., B. H. TURNER, and P. TEITELBAUM, 1971. Sensory neglect produced by lateral hypothalamic damage. *Science* 174:523–525.

MORGANE, P. J., 1961. Alterations in feeding and drinking behavior of rats with lesions in globi pallidi. *Amer. J. Physiol.* 201:420–428.

NAUTA, W. J. H., and W. R. MEHLER, 1969. Fiber connections of the basal ganglia. In *Psychotropic Drugs and Dysfunction of the Basal Ganglia.* U.S. Public Health Service Publication No. 1938, pp. 68–72.

OLTMANS, G. A., and J. A. HARVEY, 1972. LH syndrome and brain catecholamine levels after lesions of the nigrostriatal bundle. *Physiol. Behav.* 8:69–78.

POSCHEL, B. P. H., and F. W. NIUTEMAN, 1963. Norepinephrine: A possible excitatory neurohormone of the reward system. *Life Sci.* 2:782–788.

RANDRUP, A., and I. MUNKVAD, 1966. Role of catecholamines in the amphetamine excitatory response. *Nature (Lond.)* 211: 540.

RANDRUP, A., and I. MUNKVAD, 1970. Biochemical, anatomical and psychological investigations of stereotyped behavior induced by amphetamines. In *Amphetamines and Related Compounds*, E. Costa and S. Garattini, eds. New York: Raven Press, pp. 695–713.

SCHOENFELD, R., and M. URETSKY, 1972. Altered response to apomorphine in 6-hydroxydopamine-treated rats. *Europ. J. Pharmacol.* 19:115–118.

SNYDER, S. H., 1972. Catecholamines in the brain as mediators of amphetamine psychosis. *Arch. Gen. Psychiat.* 27:169–179.

SOURKES, T. L., L. J. POIRIER, and P. SINGH, 1969. Biochemical-histological-neurological models of Parkinsons disease. In *Third Symposium on Parkinson's Disease*, F. J. Gillingham and I. M. L. Donaldson, eds. Edinburgh and London: E. & S. Livingstone, pp. 54–60.

STEIN, L., 1962. Effects of interaction of imipramine chlorpromazine, reserpine, and amphetamine on self-stimulation: Possible neurophysiological basis of depression. *Recent Adv. Biol. Psychiat.* 4:288–308.

STEIN, L., and C. D. WISE, 1971. Possible etiology of schizophrenia: progressive damage to the noradrenergic reward system by 6-hydroxydopamine. *Science* 171:1032–1036.

UNGERSTEDT, U., 1968. 6-Hydroxydopamine induced degeneration of central monoamine neurons. *Europ. J. Pharmacol.* 5: 107–110.

UNGERSTEDT, U., L. L. BUTCHER, S. G. BUTCHER, N.-E. ANDÉN, and K. FUXE, 1969. Direct chemical stimulation of dopaminergic mechanisms in the neostriatum of the rat. *Brain Res.* 14:461–471.

UNGERSTEDT, U., 1970. Is interruption of the nigro-striatal dopamine system producing the "lateral hypothalamus syndrome"? *Acta Physiol. Scand.* 80:35A–36A.

UNGERSTEDT, U., and G. ARBUTHNOTT, 1970. Quantitative recording of rotational behavior in rats after 6-hydroxy-dopamine lesions of the nigro-striatal dopamine system. *Brain Res.* 24:485–493.

UNGERSTEDT, U., 1971a. Stereotaxic mapping of the monoamine pathways in the rat brain. *Acta Physiol. Scand.* (Suppl. 367) 82:1–48.

UNGERSTEDT, U., 1971b. Histochemical studies on the effects of intracerebral and intraventricular injections of 6-hydroxy-dopamine on monoamine neurons in the rat brain. In *6-Hydroxydopamine and catecholamine neurons*, T. Malmfors and H. Thoenen, eds. Amsterdam: North-Holland Comp., pp. 101–127.

UNGERSTEDT, U., 1971c. Striatal dopamine release after amphetamine or nerve degeneration revealed by rotational behavior. *Acta Physiol. Scand.* (Suppl. 367) 82:49–68.

UNGERSTEDT, U., 1971d. Postsynaptic supersensitivity after 6-hydroxydopamine induced degeneration of the nigro-striatal dopamine system in the rat brain. *Acta Physiol. Scand.* (Suppl. 367) 82:69–93.

UNGERSTEDT, U., 1971e. Mechanism of action of L-DOPA studied in an experimental Parkinson model. In *Monoamines and the Central Grey Nuclei*, J. de Ajuriaguerra, ed. Proceedings of the IV Bel-Air Symposium, pp. 165–170.

UNGERSTEDT, U., 1971f. Adipsia and aphagia after 6-hydroxy-dopamine induced degeneration of the nigro-striatal dopamine system in the rat brain. *Acta Physiol. Scand.* (Suppl. 367) 82:95–122.

UNGERSTEDT, U., 1973. Selective lesions of central monoamine pathways: Application in functional studies. In *Neurosciences Research*, S. Ehrenpreis and I. Kopin, eds. New York: Academic Press, (in press).

ZIGMOND, M. J., and E. M. STRICKER, 1972. Deficits in feeding behavior following intraventricular injection of 6-hydroxy-dopamine in rats. *Science*, in press.

6-Hydroxydopamine: A Chemical Lesion Technique for Studying the Role of Amine Neurotransmitters in Behavior

SUSAN D. IVERSEN

ABSTRACT Histochemical evidence indicates that the amine transmitters, norepinephrine (NE) and dopamine (DA), are localized in specific pathways in the CNS. One of these pathways arises from neurons in the substantia nigra (SN) and projects via the medial forebrain bundle and hypothalamus to the striatum; this pathway contains a large proportion of forebrain DA. Surgical lesions to the SN alter motor behaviors and locomotor and stereotyped responses to amphetamine and lead to severe aphagia and adipsia. These results, together with pharmacological evidence, suggest that this DA-containing system is important for normal motor behavior and feeding.

6-Hydroxydopamine (6-OHDA) is a chemical analog of DA that can be used as a valuable new tool for inducing chemically specific lesions in aminergic pathways in the CNS. When 6-OHDA is injected into the brain, it is taken up selectively by amine-containing neurons and causes degenerative changes. If the drug is applied by microinjections through cannulae into local areas of the CNS, lesions can be produced in specific aminergic pathways. Our earlier work on SN lesions has now been extended and repeated using the 6-OHDA technique. Bilateral injections of 6-OHDA into the rat SN do not produce the bizarre motor behavior seen after surgical lesions, nor did 6-OHDA change the characteristic locomotor responses to amphetamine. After 6-OHDA injections striatal DA was depleted by approximately 95%, indicating that DA is not critically involved in the control of locomotor activity. In contrast, the stereotype responses normally elicited by amphetamine were abolished in 6-OHDA lesioned animals, confirming much other evidence that suggests that DA release in the striatum is responsible for the stereotyped behavior produced by amphetamine.

THE DISCOVERY OF chemical neurotransmission in the brain is now legendary. More recently the demonstration that several different transmitters exist in the brain, not diffusely distributed but strictly localized in different anatomical systems, encouraged neuropsychologists to investigate the functional significance of these transmitter systems for the control of behavior. This chapter is not presented simply to demonstrate that a general loss of neurotransmitters abolishes behavior but to discover the importance of the different transmitters and the different parts of their anatomical distribution for the complex fabric of behavior. This is not a simple task for the complexity of the brain is rivaled only by the intricacies of behavior. However, a wealth of methodology is available to apply to these problems, ranging from the well-tried lesion and stimulation techniques of neuropsychology to the extensive pharmacological knowledge of the interaction between drugs and neurotransmitters.

The nigrostriatal pathway and motor behavior

Dopamine (DA) has been identified as a transmitter and is localized to an anatomical system different from that containing the related amine transmitter norepinephrine (NE) (Dahlström and Fuxe, 1964; Ungerstedt, 1971a). Within this framework investigation has been undertaken of the role of the dopamine-containing pathway between the substantia nigra and the corpus striatum in some motor behaviors. Principally, interest has focused on the increased locomotor activity and stereotypy (repetitive behavior including sniffing, rearing, neck movements, and gnawing) seen with varying doses of the stimulant drug amphetamine, and the hypothesis that the nigrostriatal pathway mediates these behavioral effects of the drug. Despite the limited nature of the problem and the simplicity of the behaviors, the project has some important properties that make it attractive from an experimental point of view.

1. The pathway between the nigra and the corpus striatum contains 75 to 80% of forebrain DA. In percentage terms it is a giant neurochemical pathway, although with classical anatomical techniques it is barely visible.

2. It is physically accessible for experimental manipulations. The cell bodies in the substantia nigra (SN) are separated from the axons of the dorsal and ventral NE bundles at the level of the SN. Furthermore, unlike, for example, hypothalamic nuclei, the SN is not closely

SUSAN D. IVERSEN Department of Experimental Psychology, University of Cambridge, England surrounded by pathways or structures that, if damaged accidentally, produce behavioral deficits likely to confuse the results.

3. There is extensive pharmacological literature on drugs which influence the synthesis, release, uptake, and receptor-stimulating properties of DA.

4. There is in vitro and in vivo evidence that amphetamine interacts with amine transmitters in the brain.

5. The fact that amphetamine-induced stereotypy and activity are motor behaviors and that the drug interacts with DA mechanisms provides the starting hypothesis that the nigrostriatal pathway mediates these behaviors.

Studies investigating the role of the nigrostriatal pathway for locomotor activity and stereotyped behavior

Effect of 6-Hydroxydopamine Lesions to the Substantia Nigra 6-Hydroxydopamine (6-OHDA) is a chemical analog of the amine transmitters NE and DA, which if injected intraventricularly is taken up by and destroys amine-containing neurons, thus depleting NE and DA in all brain regions (Uretsky and Iversen, 1970 Iversen and Uretsky, 1971). Ungerstedt (1971b) advanced the value of 6-OHDA as a chemical lesion technique by the use of implanted cannulae for local application to brain structures. When 6-OHDA is injected (8 μg 6-OHDA in 2 μl) bilaterally through implanted cannulae into the SN, substantial depletion of striatal DA can be achieved (Table I) following the drug induced damage and degeneration of DA-containing neurons (Creese and Iversen, 1972). Sprague Dawley rats were given such lesions or appropriate sham operations and at 4, 11, and 19 days after operation were tested in photocell cages immediately after a dose of 1.5 mg/kg *d*-amphetamine. The results are shown in Figure 1. In controls 1.5 mg/kg *d*-amphetamine characteristically produced stimulation of locomotor activity reaching a maximum at 1 hr after treatment. This was followed by the emergence of sterotyped behavior including sniffing, rearing, and neck movements. The balance, intensity and the time course of these two behaviors change depending on the dose of amphetamine. While photocell

TABLE I

Biochemical assessment of damage to catecholamine pathways in rat brain after various lesions, and the effects on amphetamine induced behavioral responses. Results are given as remaining catecholamine (DA in striatum, NE in hypothalamus and cortex) or tyrosine hydroxylase activity (marked *) as % values in untreated control animals

Lesion	Striatum DA	Hypo-thalamus NE	Cortex NE	Amphetamine Responses Locomotor Activity	Stereotypy	Reference
6-OHDA lesion to s. nigra (*nonaphagic*)	10*	57*	—	√ enhanced	×	a
6-OHDA lesion to s. nigra (*aphagic*)	4	27	60	—	—	b
6-OHDA lesion to ventral bundle	67	21	—	√ enhanced	√	c
6-OHDA lesion to dorsal bundle	92	44	—	√	√	c
Intracerebral 6-OHDA (neonates)	0–2*	36*	0*	×	×	d
Intraventricular 6-OHDA (adults—whole brain assay)						
2 × 250 μg	15	20	4	√	√	e
1 × 100 μg	65	38	—	√ enhanced	√	e
Surgical lesions to s. nigra	64	73	—	√	√	f

*Tyrosine hydroxylase activity.
—Not measured.
[a] Creese and Iversen, 1972.
[b] Evetts et al. (unpublished).
[c] Creese (unpublished).
[d] Creese (unpublished).
[e] Evetts et al., 1970.
[f] Iversen, 1971.

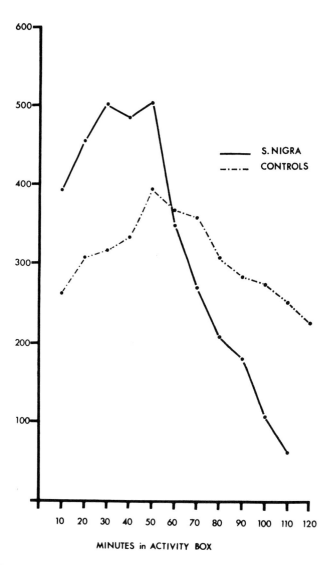

MINUTES in ACTIVITY BOX

FIGURE 1 The mean locomotor and stereotypy response to 1.5 mg/kg *d*-amphetamine in rats with bilateral 6-OHDA lesions to the substantia nigra and sham operated controls. The responses were averaged for three test sessions on different days and the saline response subtracted.

accurately record locomotor activity, their use in measuring stereotypy demands caution. Stereotyped behavior may yield a high or a low photocell count depending on the position of the animal relative to the beams when intense stereotypy is occurring or depending on the amount of locomotor activity occurring in conjunction with the stereotypy. Therefore, when estimates of stereotypy are made, the photocell record is supported by direct observation of the rat.

In 6-OHDA treated animals, the locomotor response to amphetamine during the first hour was significantly enhanced (Figure 1). It is important to mention that bilateral surgical (Iversen, 1971) or 6-OHDA

(Ungerstedt, 1971c) lesions to the substantia nigra result in hypodipsia and hypophagia of varying severity. Deprivation has been reported to enhance the locomotor response to amphetamine (Campbell and Fibiger, 1971), and it seemed possible that this was the explanation for the enhanced amphetamine response during the first test hour in the present experiments. Accordingly, the sham-operated animals were deprived to the same body weight as the 6-OHDA nigral animals (Creese and Iversen, 1972), and the enhanced amphetamine response was still present. More recently Creese has controlled for the weight difference between the experimental and control groups by depriving both groups of food and water for 5 days after the operation. At this time the weights of the two groups were equal and were subsequently maintained by tube feeding. Again an enhanced locomotor response to amphetamine was seen in the animals with nigral lesions, thus discounting weight loss as the direct cause of the enhanced amphetamine responses.

By contrast, stereotyped behavior did not emerge during the second test hour in the animals with 6-OHDA lesions (Figure 1). The failure of amphetamine to induce stereotypy in the 6-OHDA treated animals was verified by treating the rats with a drug combination (150 mg/kg iproniazid + 5 mg/kg *d*-amphetamine), which elicited intense stereotyped behavior in control animals. The animals treated in this way were rated for stereotypy by recording beam interruptions and by direct observation for 3.6 hr after injection. It was confirmed that stereotyped responding was blocked in the rats with lesions in the SN.

This finding, together with previously published evidence that pharmacological manipulation of dopaminergic mechanisms (in some cases directly in the corpus striatum) elicits stereotypy, confirms the central role of the nigro-striatal DA system for amphetamine-induced stereotyped behavior (Munkvad et al., 1968).

The failure of 6-OHDA lesions to the SN to block the locomotor response to amphetamine might be cited in support of the suggestion that the NE containing pathways of the forebrain are principally involved in this behavior (Taylor and Snyder, 1971). However, in recent experiments the effects of amphetamine in rats subjected to bilateral 6-OHDA lesions to either the ventral *or* dorsal noradrenergic bundles at the level of the midbrain have been studied. Such lesions depleted forebrain NE to 40% or less of control values (Table I; in some animals to very low levels), and yet the amphetamine locomotor response was normal, indeed even enhanced after the ventral bundle lesion. The study is not yet complete, because we do not at present have data on groups of animals with a loss of hypothalamic NE as complete as the loss of striatal DA seen after SN lesions, but the preliminary results suggest that depletion of NE alone, like DA alone, will not block the locomotor response.

It may be asked therefore if any manipulation of the NE or DA systems can abolish the locomotor response to amphetamine.

EFFECT ON AMPHETAMINE-INDUCED MOTOR BEHAVIOR OF INHIBITING AMINE SYNTHESIS IN THE BRAIN

L-α-Methyl-*p*-tyrosine (αMT) prevents the synthesis of both DA and NE by inhibiting tyrosine hydroxylase, the rate limiting enzyme on the pathway from tyrosine to DOPA. Two hours after an i.p. injection of αMT (200 mg/kg), mice no longer showed a locomotor response to 4 mg/kg *d*-amphetamine (Figure 2) in agreement with previously reported findings (Rech et al., 1966; Weissman et al., 1966; and Stolk and Rech, 1970).

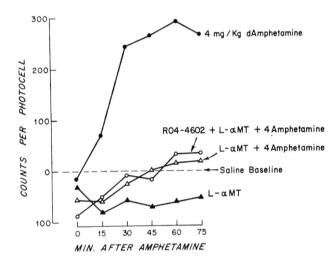

FIGURE 2   The effect of 200 mg/kg of α-methyl tyrosine (αMT) on the locomotor response to 4 mg/kg *d*-amphetamine in mice. The effect of αMT alone and pretreatment with the dopa decarboxylase inhibitor Ro4-4602 are also illustrated. The saline and αMT baselines have been subtracted where appropriate.

The possibility existed that αMT might abolish the amphetamine effects because of a receptor blocking effect of α-methyl-tyramine formed by decarboxylation from αMT in vivo. This interpretation is discounted, however, by the finding that after treatment with the DOPA decarboxylase inhibitor Ro4-4602, the αMT still blocked the amphetamine responses. Ro4-4602 inhibits the synthesis of DA to NE and was used in this experiment to prevent the conversion of αMT to α-methyl-tyramine, a "false transmitter."

The well-established effect of αMT in antagonizing the effects of amphetamine clearly implicates the catecholamines in the locomotor response to amphetamine, and the

task is to corroborate this result using alternative ways to deplete NE and DA.

THE USE OF 6-OHDA TO ACHIEVE NEAR TOTAL NE AND DA LOSS

It should be possible with multiple placement and repeated injections of 6-OHDA to achieve virtually total depletion of both DA and NE in the adult rat. Unfortunately, the severe aphagia and adipsia associated with successful SN lesions would render any such preparation undesirable in behavioral terms, irrespective of any additional effects due to the NE loss.

An alternative way of achieving such severe catecholamine depletion involves intracerebral injections of 6-OHDA (100 μg free base in 10 μl) into neonatal rats. Rats treated in this way on days 5, 7, and 9 are about half the weight of control animals at 3 months of age, have abnormal feeding and learning behavior, and a range of endocrinological defects (Breese and Traylor, 1972). Creese has recently found that at three months of age such animals show a totally blocked response to amphetamine both with respect to induced locomotor activity and stereotypy (Figures 3 and 4).

The analysis of the pharmacological and behavioral properties of these animals is at an early state but it is already clear that, despite the lack of presynaptic amines, the amine receptors appear to be functional. Indeed, if apomorphine was used to induce stereotypy (by direct stimulation of dopamine receptors), a dose of 1 mg/kg which was ineffective in normal animals, produced intense stereotyped behavior in the animals treated with 6-OHDA as neonates (Figure 4).

Preliminary assay data on these animals (Table I) indicated that DA was almost completely absent from the

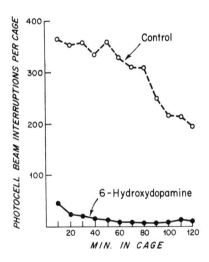

FIGURE 3   The blocked locomotor response to 1.5 mg/kg *d*-amphetamine in rats treated as neonates with intracerebral 6-OHDA.

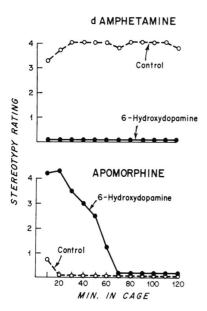

FIGURE 4 *Upper:* The blocked stereotypy to 5 mg/kg *d*-amphetamine in rats treated as neonates with intracerebral 6-OHDA. *Lower:* The enhanced stereotypy induced by 1 mg/kg apomorphine in rats treated as neonates with intracerebral 6-OHDA.

striatum, the dorsal NE system was also severely depleted, and the ventral NE system to the hypothalamus drastically reduced. The resistance of the ventral NE bundle even to 6-OHDA injections in neonatal animals supports the view that amine neurons vary in their sensitivity to 6-OHDA and probably also explains why after peripheral 6-OHDA injections in the neonate, Clark et al. (1972) found dorsal but not ventral bundle NE depletion.

Synthesis of the results

In summary, it has been shown that the dopaminergic nigro-striatal pathway plays a central role in amphetamine induced stereotyped behavior. Although, as Ahlenius and Engel (1972) have suggested, the NE system may modulate the stereotypy mechanism via an inhibitory influence, it cannot mediate the behavioral response in the absence of the DA system.

However, both NE and DA appear to be critically involved in the locomotor response to amphetamine, as virtually total depletion of either system does not by itself block the behavior, whereas depletion of both amines does. Such depletion must apparently be near total, and on the basis of the preliminary experiment on neonatally treated animals, the dorsal NE system may be more important than the ventral bundle. The projection of the dorsal NE system to the hippocampus and cortex is consistent with a role in behavioral arousal.

Ungerstedt (this volume) maintains that DA plays the central role in amphetamine induced locomotor stimulation, because he has demonstrated a block of such responses in adult rats after 6-OHDA SN lesions that produces near total striatal DA loss. Unfortunately, however, it seems that a "successful" SN lesion always results in substantial forebrain NE loss, presumably because of damage to the NE axons in the vicinity of the SN. Regional amine assay data on the brains of such animals supports this view (Table II); although the degree of DA depletion correlates with the block of the locomotor response and the severity of aphagia and adipsia, so does the associated NE depletion.

TABLE II

DA and NE levels in the striatum and hypothalamus/limbic brain regions after 6-OHDA lesions to the substantia nigra. These animals ($N = 5$) were severely aphagic and adipsic

	DA in Striatum	NE in Hypothalamus/ Limbic System
	7.6	30
	8.0	60
	1.0	14
	5.0	15
	0	15
Mean	4.3 ± 1.6	27 ± 9.0

The strongest evidence against a central role for DA in both locomotor activity and stereotypy responses comes from the present results on animals with 90% striatal DA loss. Such DA depletion blocked stereotypy and not locomotor activity. A 10% residue of dopamine plus supersensitive, denervated receptors might still account for the normal locomotor response. However, special pleading is required if such a pool of dopamine could maintain one behavior (locomotor activity) but not a closely related one (stereotypy), for which there is much stronger evidence of DA involvement.

The present experiments, together with accumulating evidence from studies using inhibitors of enzymes involved in amine synthesis to deplete selectively DA or NE (Svensson and Waldeck, 1970), support the view that DA and NE interact in the control of amphetamine induced locomotor activity. In various experiments with 6-OHDA administered either locally to amine pathways or intraventricularly (Evetts et al., 1970), we have seen an enhancement of the locomotor responses to amphetamine. In all of these cases, 6-OHDA treatment resulted in an unbalanced depletion of DA and NE (Table I). If NE synthesis is temporarily inhibited with the dopamine-β-hydroxylase inhibitor, FLA-63, a depression of spontaneous locomotor activity is observed (Svensson, 1971), suggesting a normal facilitatory role of NE for this behavior, compared with its inhibitory influence on stereotypy (Ahlenius and Engel, 1971).

On this note of dynamic interplay between amine systems, it is relevant to return to one of the early experiments of this project. Before 6-OHDA became available, electrolytic lesions were used to damage the SN and induce neuronal degeneration and DA loss. Within hours of recovery from the anesthetic, intense stereotyped running, neck movements, and gnawing were seen in the absence of any drug treatment, and these spontaneous behaviors gradually faded to control levels over the next 10 days. The time course of these behaviors follows closely the release of striatal dopamine consequent upon neuron terminal degeneration in the corpus striatum (Faull and Laverty, 1969). It is tempting to speculate that these immediate postoperative effects on behavior reflect the sudden changes in the amine balance in the forebrain, consequent upon degeneration and release of endogenous amines. It may be likened to the stereotypy induced by temporarily inhibiting the synthesis of NE with the dopamine-β-hydroxylase inhibitor FLA-63 (Ahlenius and Engel, 1972).

Fourteen days after surgery, the response to amphetamine was tested, and in some animals the locomotor response to amphetamine was virtually abolished (Simpson and Iversen, 1971). Even in animals where such a block was not seen, the characteristic patterns of interaction between locomotor activity and stereotypy, seen at different doses of amphetamine, were disrupted. This result was puzzling because the striatal DA depletion was rarely more than 50%, and there was no obvious correlation between the completeness of the blocked amphetamine locomotor response and the degree of DA depletion (Iversen, 1971). The 6-OHDA results confirm the suspicion that the results of surgical lesions reflect *not* striatal DA depletion but surgical interference with extrapyramidal motor systems. Stereotypy was clearly not abolished in these rats with electrolytic SN lesions. It now seems likely that because the locomotor response was disrupted the normal balance between the behaviors was distorted, and stereotypy could emerge without competition from locomotor activity and with the same time course, irrespective of the dose of amphetamine (Iversen, 1971).

However, while the permanent surgical disruption is sufficient to distort the balance between the locomotor and stereotypy responses to amphetamine, it is not sufficient either in depletion or surgical terms to block amphetamine induced stereotypy. This suggests that while locomotor activity and stereotypy may be mutually dependent on the same NE and DA pools, they may find their specificity through NE and DA interactions operating on different efferent systems.

REFERENCES

AHLENIUS, S., and J. ENGEL, 1971. Behavioral and biochemical effects of L-dopa after inhibition of dopamine-β-hydroxylase in reserpine pretreated rats. *Naunyn Schmeidebergs Arch. Pharmakol.* 270:349–360.

AHLENIUS, S., and J. ENGEL, 1972. Effects of a dopamine (DA)-β-hydroxylase inhibitor on timing behavior. *Psychopharmacologia* 24:243–246.

BREESE, G. R., and T. D. TRAYLOR, 1972. Developmental characteristics of brain catecholamines and tyrosine hydroxylase in the rat: Effects of 6-hydroxydopamine. *Br. J. Pharmacol.* 44:210–222.

CAMPBELL, B. A., and H. C. FIBIGER, 1971. Potentiation of amphetamine-induced arousal by starvation. *Nature (Lond.)* 233:424–425.

CLARK, D. N. J., R. LAVERTY, and E. L. PHELAN, 1972. Long-lasting peripheral and central effects of 6-hydroxydopamine in rats. *Br. J. Pharmacol.* 44:831–842.

CREESE, I., and S. D. IVERSEN, 1972. Amphetamine response in rat after dopamine neuron destruction. *Nature New Biol.* 238:247–248.

DAHLSTRÖM, A., and K. FUXE, 1964. Evidence for the existence of monoamine-containing neurons in the central nervous system. *Acta Physiol. Scand.* (Suppl.) 232:1–55.

EVETTS, K. D., N. J. URETSKY, L. L. IVERSEN, and S. D. IVERSEN, 1970. Effects of 6-hydroxydopamine on CNS catecholamines, spontaneous motor activity and amphetamine induced hyperactivity in rats. *Nature (Lond.)* 225:961–962.

FAULL, R. L. M., and R. LAVERTY, 1969. Changes in dopamine levels in the corpus striatum following lesions in the substantia nigra. *Exp. Neurol.* 23:332–340.

IVERSEN, L. L., and N. J. URETSKY, 1971. Biochemical effect of 6-hydroxydopamine on catecholamine containing neurons in the rat central nervous system. In *6-Hydroxydopamine and Catecholamine Neurons*, T. Malmfors and H. Theonen, eds. Amsterdam: North-Holland Publishing, pp. 171–186.

IVERSEN, S. D., 1971. The effect of surgical lesions to frontal cortex and substantia nigra on amphetamine responses in rats. *Brain Res.* 31:295–311.

MUNKVAD, I., H. PAKKENBERG, and A. RANDRUP, 1968. Aminergic systems in basal ganglia associated with stereotyped hyperactive behavior and catalepsy. *Brain Behav. Evol.* 1:89–100.

RECH, R. H., H. K. BORYS, and K. E. MOORE, 1966. Alterations in behavior and brain catecholamine levels in rats treated with α-methyltyrosine. *J. Pharmac. Exp. Ther.* 153:412–419.

SIMPSON, B. A., and IVERSEN, S. D., 1971. Effects of substantia nigra lesions on the locomotor and stereotypy responses to amphetamine. *Nature, New Biol.* 230:30–32.

STOLK, J. M., and R. H. RECH, 1970. Antagonism of D-amphetamine by alpha-methyl-L-tyrosine: Behavioral evidence for the participation of catecholamine stores and synthesis in the amphetamine stimulant response. *Neuropharmacology* 9:249–263.

SVENSSON, T. H., 1971. On the role of central norepinephrine in the regulation of motor activity and body temperature in the mouse. *Naunyn Schmeidebergs Arch. Pharmakol.* 271:111–120.

SVENSSON, T. H., and B. WALDECK, 1970. On the role of brain catecholamines in motor activity: Experiments with inhibitors of synthesis and of monoamine oxidase. *Psychopharmacologia* 18:357–365.

TAYLOR, K. H., and S. H. SNYDER, 1971. Differential effects of D- and L-amphetamine on behavior and on catecholamine disposition in dopamine norepinephrine containing neurons of rat brain. *Brain Res.* 28:295–309.

UNGERSTEDT, U., 1971a. Stereotaxic mapping of the monoamine pathway in the rat brain. *Acta Physiol. Scand. (Suppl.)* 367:1–48.

UNGERSTEDT, U., 1971b. Histochemical studies on the effect of intracerebral and intraventricular injections of 6-hydroxydopamine on monoamine neurons in the rat brain. In *6-Hydroxydopamine and Catecholamine Neurons*, T. Malmfors and H. Theonen, eds. Amsterdam: North-Holland Publishing, pp. 101–127.

UNGERSTEDT, U., 1971c. Adipsia and aphagia after 6-hydroxydopamine induced degeneration of the nigrostriatal dopamine system. *Acta. Physiol. Scand.* (Suppl.) 367:95–122.

URETSKY, N. J., and L. L. IVERSEN, 1970. Effects of 6-hydroxydopamine on catecholamine containing neurons in the rat brain. *J. Neurochem.* 17:269–278.

WEISSMAN, A., B. K. KOE, and S. J. THENEN, 1966. Antiamphetamine effects following inhibition of tyrosine hydroxylase. *J. Pharm. Exp. Ther.* 151:339–352.

52 Adrenergic Receptor Mechanisms in Eating and Drinking

SARAH FRYER LEIBOWITZ

ABSTRACT The role of central adrenergic receptor mechanisms in the normal regulation of food and water intake is the primary subject. Experimental evidence from studies using central drug injections is presented. Several topics are discussed, including (a) the function of the two types of adrenergic receptors, alpha and beta, (b) the localization of the different receptor effects in different hypothalamic sites, and (c) the central mode of action of the drugs amphetamine and chlorpromazine.

PHARMACOLOGICAL studies in the peripheral nervous system have demonstrated the important role that adrenergic mechanisms play in controlling the functions of peripheral tissues. Detailed analyses of these peripheral adrenergic mechanisms, using several different adrenergic stimulants and receptor blockers, have suggested that there are at least two types of adrenergic receptors, which Ahlquist (1948) called alpha and beta. Stimulation of these two types of receptors has sometimes been found to produce diametrically opposite effects, such as on vascular muscle where alpha-receptor activity produces constriction and beta-receptor activity produces dilatation. In other cases, alpha- and beta-receptor stimulation may have synergistic effects, such as on the intestinal smooth muscle where stimulation of either type of receptor causes relaxation.

The various effects of adrenergic receptor stimulation have been studied with the help of the three agonists: norepinephrine (NE), epinephrine (EPI), and isoproterenol (ISOP), which differ in the ratio of their effectiveness in stimulating peripheral alpha and beta receptors (Innes and Nickerson, 1970). Norepinephrine has potent alpha-receptor action but somewhat weaker action on beta receptors. Isoproterenol, in contrast, acts predominantly on beta receptors and has very little alpha-receptor activity. Finally, EPI has potent action on both alpha and beta receptors.

During the past decade or so, some pharmacological studies have been carried out in the central nervous system that suggest that adrenergic receptor mechanisms also exist in the brain and that they may regulate, possibly through the transmitter action of NE, specific behavioral and physiological responses. Histochemical and biochemical studies have demonstrated the existence in the brain of NE and also of the enzyme that converts NE to EPI. Direct applications to the brain of these agents, or of compounds that affect their level or activity in the brain, have been found to alter central neural activity and, ultimately, behavior. The role of central adrenergic receptor mechanisms in the control of ingestive behavior has been extensively investigated in the rat, and the evidence to be described here suggests that there are alpha- and beta-adrenergic receptors in the brain which modulate (that is, elicit and inhibit) the ingestion of food and water.

Technique for central drug injection

To study the relationship between central adrenergic mechanisms and behavior, a technique has been used that allows drugs to be repeatedly administered into the brains of unanesthetized, unrestrained subjects. This technique involves the use of a cannula, usually a 23- to 29-gauge hypodermic needle, which is stereotaxically implanted into a specific area of the brain of an anesthetized subject and then fixed in place by an acrylic cement on top of the skull. After the subject has recovered from surgery, drugs are administered through this cannula, either in crystalline form or in solution (in volumes of 0.1 to 1.0 μl) and, immediately thereafter, the responses of the unrestrained subject are observed. The procedures used for examining drug effects on the ingestive responses, feeding and drinking, are generally quite straightforward. In some experiments, the subjects are tested in a satiated state, in which case they are given food and water ad libitum before the test. In other experiments, the subjects are tested in a deprived state, in which case they are deprived of either food or water for a certain period of time before the test. During the test, measured food and/or water is made available, and the subject's consumption of either is recorded at frequent intervals.

SARAH FRYER LEIBOWITZ   The Rockefeller University, New York, N.Y.

Alpha-adrenergic receptors that elicit feeding

A connection between central adrenergic systems and the elicitation of feeding behavior was first demonstrated in the rat by Grossman (1962a), who found that injection of NE or EPI directly into the hypothalamus of satiated subjects induced eating (Table I). Further examination of this adrenergic feeding effect has shown that its magnitude is dependent upon the dose of the agonist (Miller et al., 1964; Booth, 1968; Leibowitz, in preparation) and that it can be obtained at several diencephalic sites (Booth, 1967; Coury, 1967).

One especially sensitive site appears to be the "perifornical" hypothalamus, an area near the fornix, the caudal half of the anterior hypothalamus. At this site, a dose of EPI at least as low as 0.25 nmole can produce a reliable feeding response (Leibowitz, 1972d). The latency of the adrenergically elicited feeding response, which can vary from less than a minute to as much as 10 min, tends to decrease with increase in dose and also appears to be affected by the presence or absence of drinking water. That is, when injected with NE or EPI, a rat frequently drinks some water (see below) before starting to eat at 5 to 6 min after injection. If water is not available, the feeding response occurs sooner, usually 2 to 3 min after injection. The feeding itself generally lasts between 10 and 30 min. However, the stimulating effect on adrenergic receptors may remain active for well over an hour, as shown for example by animals who exhibit enhanced eating when given food for the first time as much as 60 min after injection.

Studies designed to determine which type of adrenergic receptor, alpha or beta, is mediating this stimulating effect on food intake suggest that it is alpha in nature. Alpha-receptor blocking agents, but not beta-receptor blocking agents, can reliably reduce or even abolish the enhancement of eating induced by hypothalamic injection of NE or EPI in either satiated or hungry rats (Grossman, 1962b; Slangen and Miller, 1969; Leibowitz, in preparation). Evidence to suggest that central alpha-adrenergic receptors are physiologically active during natural hunger is provided by Grossman (1962b) and Leibowitz (1970c), who showed that eating by a hungry rat can be reliably suppressed by peripheral or central injection of an alpha-receptor blocker. Further evidence is provided by the study of Slangen and Miller (1969), which showed that eating can be induced in satiated rats by a combination of drugs which presumably causes an overflow of endogenous NE into the synaptic cleft.

In addition to these studies in the rat, there are studies in the monkey that show that hypothalamic injection of NE can also increase food consumption in this species

TABLE I

Alpha- and beta-adrenergic effects on ingestive behavior induced by hypothalamic injection of epinephrine or norepinephrine

Ingestive Behavior	Adrenergic Effect	Type of Receptor*	Drug Dose (nmole)	Hypothalamic Site of Injection	Latency (min)	Magnitude (increases with dose)	Duration (min) (increases with dose)	Special Characteristics
Feeding	Stimulation	α	0.5	Perifornical ventromedial	3–5	Increase of 1–6 g (baseline = 0.3 g)	at least 90**	Often preceded by drinking; eating is vigorous and continuous
	Suppression	β	10.0	Perifornical lateral	0–1	Suppression of up to 90%	at least 120	Antagonized by above alpha-adrenergic feeding effect; greater suppression observed in the presence of an alpha-receptor blocker
Drinking	Suppression	α	0.005	Perifornical ventromedial	3–5	Suppression of up to 90%	at least 180	Suppresses drinking induced by deprivation, hyperosmolarity, hypovolemia, and central drug injections
	Stimulation	α and β	0.5	Perifornical	1–2	Increase of 1–6 ml (baseline = 0.0 ml)	1–5**	Brief but vigorous; associated with alpha-adrenergic feeding effect

*The type of receptor mediating the effect is determined on the basis of the blocking effects of alpha- and beta-receptor blockers. An alpha-receptor effect is blocked by the alpha blockers, phentolamine and tolazoline. A beta-receptor effect is blocked by the beta blockers, propranolol, MJ 1999 and LB-46.

**Duration of the alpha feeding-stimulation effect and the alpha plus beta drinking-stimulation effect is determined by giving the subjects food and water, respectively, at variable intervals after injection.

(Myers, 1969). This evidence suggests that in the monkey, as in the rat, central adrenergic mechanisms may play a role in controlling feeding behavior.

Beta-adrenergic receptors that inhibit feeding

In contrast to the two agonists NE and EPI, which have alpha-receptor activity, the relatively pure beta-receptor agonist ISOP fails to elicit feeding in the rat (Myers and Yaksh, 1968; Slangen and Miller, 1969; Leibowitz, 1972c). This evidence and the failure of beta-receptor blockers to suppress feeding elicited by NE or EPI indicate that in the rat central beta-adrenergic receptors, unlike alpha-adrenergic receptors, are not important for eliciting the eating response. However, there is some evidence that suggests that central beta receptors may indeed influence feeding behavior but in the opposite direction (Table I) (Leibowitz, 1970a; Goldman et al., 1971). That is, in contrast to the stimulating effect of alpha-receptor activity, beta-receptor activity appears to have a strong inhibitory effect on eating behavior. This suppression of feeding was first observed with peripheral administration of the adrenergic agonists (Miller, 1965; Conte et al., 1968; Russek et al., 1968) and was subsequently observed and analyzed by Leibowitz with central administration of drugs.

Perifornical hypothalamic injection of the beta agonist ISOP produces, immediately after injection in hungry rats, a strong suppression of feeding, the magnitude and duration of which is dependent upon dose (Leibowitz, 1970a, and unpublished work). At a dose of 5 nmoles, ISOP reliably suppresses feeding for 15 to 30 min, and at 20 times this dose, the suppression lasts for over 2 hr. This effect of ISOP is totally eliminated by preceding injections of beta-receptor blockers but not alpha-receptor blockers.

Perifornical hypothalamic injection of EPI, an agonist with potent beta-receptor action as well as potent alpha-receptor action, can have, in addition to its alpha-receptor potentiation of feeding effect (see above), a strong suppressing effect on food intake that is blocked by beta-receptor blockers (Leibowitz, 1970a). However, especially at relatively low doses, this beta-receptor suppression of feeding effect may not always be reliably apparent, because of EPI's simultaneous alpha-receptor action, and in some cases it may be exhibited only in the presence of an alpha-receptor blocker (Leibowitz, in preparation). These studies of the effects of adrenergic blocking agents on EPI-induced changes in food intake demonstrate the ability of centrally injected EPI to stimulate both alpha and beta receptors and lend further support to the suggestion that these two types of receptors have antagonistic control of feeding behavior.

In the peripheral nervous system, the beta-receptor action of EPI is generally found to be more potent than that of NE. In the brain this also appears to be the case, at least with respect to the agonists' stimulation of the beta receptors that inhibit food consumption. Injection of NE alone into the perifornical hypothalamus of hungry rats is most often found to produce the alpha-receptor effect, a potentiation of food consumption. However, when injected with an alpha-receptor blocker, a marked beta suppressing effect of this agonist on feeding can be observed (Leibowitz, in preparation). These findings indicate that NE can activate both the alpha receptors for feeding and the beta receptors for inhibition of feeding, and therefore that endogenous NE might be a neurotransmitter in the central adrenergic system that has diametrically opposite effects on feeding behavior. The feeding stimulation effect that can be produced by a high dose of a beta-receptor blocker and the feeding suppression effect that can be produced by a high dose of an alpha-receptor blocker (Leibowitz, 1970c) lend further support to the suggestion that these two types of adrenergic receptor systems are physiologically active during normal regulation of feeding.

Alpha-adrenergic receptors that suppress drinking

The above findings demonstrate the importance of central adrenergic mechanisms in the control of feeding behavior. Additional evidence, which is now described, suggests that central adrenergic mechanisms may also play a role in the control of drinking behavior (Table I).

Hypothalamic injection of NE was first reported by Grossman (1962a) to have a suppressing effect upon water consumption in thirsty rats. This effect was further analyzed by Leibowitz (1971a, 1972b, in preparation) and was found to be produced by perifornical hypothalamic injection of either NE or EPI but not of the beta agonist ISOP. This inhibitory effect of adrenergic stimulation on drinking, which appears to have a latency of up to 5 min, can be reliably observed for about 2 min at a dose as low as 0.005 nmole. As the dose is increased beyond this point, both the magnitude and duration of the effect increase.

Experiments with receptor blocking agents (Leibowitz, 1972b, in preparation) demonstrate that the suppression of water intake produced by central NE or EPI injection is blocked by alpha-receptor blockers but not by beta-receptor blockers. This finding indicates that it is alpha, and not beta, receptors that cause an inhibition of drinking. Evidence to suggest that these alpha receptors, presumably located in the brain, are physiologically active in the normal regulation of water consumption is provided by the finding that hypothalamic injection of an alpha-receptor blocker alone can stimulate drinking (Leibowitz, 1971a).

In addition to suppressing deprivation-induced drinking, hypothalamic injection of NE is found to suppress water consumption elicited by cellular dehydration, hypovolemia, and central injections of other drugs (Leibowitz, 1972b; Singer and Kelly, 1972). These findings further emphasize the importance of central adrenergic systems in the control of drinking behavior and suggest that endogenous NE may be a mediator in these systems.

Adrenergic receptors that elicit drinking

Peripheral administration of ISOP, a beta agonist, was found by Lehr et al. (1967) to produce in rats a drinking response which occurred sporadically over a period of 2 to 3 hr. Lehr suggested that ISOP was producing this response by entering the brain and stimulating central beta receptors that elicit drinking. Houpt and Epstein (1971), however, proposed an alternative hypothesis. They suggested that by causing a release of renin from the kidneys, ISOP was increasing the levels of circulating angiotensin and that it was this substance which acted directly on the brain to elicit drinking. These investigators, finding that nephrectomy abolishes drinking induced by peripheral ISOP, suggest that the kidneys are in some way, directly or indirectly, involved in this beta-adrenergic drinking phenomenon. However, it leaves unanswered the question of whether beta receptors that induce drinking do exist in the brain.

In attempting to answer this question, Leibowitz (1971a) injected ISOP directly into the lateral hypothalamus of the rat. At doses of 25 to 100 nmole, a reliable drinking response was obtained, which resembled the response observed after peripheral ISOP injection (it was sporadic but persistent). This drinking response was blocked by a centrally injected beta blocker but not by an alpha blocker. In view of the fact that the doses of ISOP used centrally were similar to the doses found to be effective when peripherally administered (Lehr et al., 1967), this evidence can only tentatively be accepted as support for Lehr's hypothesis that beta receptors for drinking do exist in the brain. Since several other hypothalamic sites tested did not respond positively to ISOP and since these negative sites were in some cases located right next to the positive lateral hypothalamic site, it would appear that ISOP injected into the lateral hypothalamus is acting locally on central beta receptors and is not simply diffusing to other central or even peripheral sites. Evidence to support this central mediation hypothesis should ideally include the locating of a site that is

sensitive to a dose of ISOP that is lower than the dose required to produce the effect when peripherally injected.

Lateral hypothalamic injection of EPI or NE does not produce the sporadic, long-lasting drinking elicited by ISOP injection. However, these two agonists, when injected into the perifornical hypothalamus of the satiated rat, do elicit a quite different drinking response (Table I). This response, which is dose-dependent, starts almost immediately after injection and lasts for only about 5 min (in contrast to the ISOP response which starts 5 to 10 min after injection and occurs sporadically over a period of 2 hr) (Leibowitz, 1971b, in preparation). This brief drinking phenomenon, which can be reliably elicited by centrally injected NE or EPI at a dose at least as low as 0.5 nmole, is not seen after peripheral administration of these agonists. Studies with centrally injected blocking agents show that it can be eliminated both by beta-receptor blockers and by alpha-receptor blockers. Interestingly, this drinking effect is followed immediately by, and correlated in magnitude with, an alpha-receptor feeding response (see above). This connection between the adrenergic drinking phenomenon and the ingestion of food suggests that the water consumption elicited by NE or EPI is a food-associated drinking response, a type of drinking that is very frequently exhibited by rats under normal conditions (Fitzsimons and Le Magnen, 1969; Kissileff, 1969). In view of the evidence that central alpha-receptor activity elicits feeding and the suggestive evidence discussed above that central beta-receptor activity may elicit drinking, it is especially interesting that both alpha and beta receptors are involved in mediating this food-associated drinking behavior.

In contrast to NE or EPI, the pure beta agonist ISOP, when injected alone into the perifornical hypothalamus, only occasionally elicits the brief drinking response described above and never elicits the feeding response that follows the drinking. It has been found, however, that when ISOP is centrally injected simultaneously with NE or EPI, this beta agonist can greatly enhance the food-associated drinking response elicited by either of the two other agonists (Leibowitz, 1972a). This result, which is not seen with peripheral ISOP injection, confirms the involvement of central beta receptors in eliciting drinking behavior, at least that drinking which is closely associated with eating.

Localization within the hypothalamus of alpha- and beta-receptor effects

Most of the experiments described above were carried out in the perifornical hypothalamus, an area that is dense with adrenergic terminals and that is found to be especially sensitive to almost all of the effects of the adrenergic agonists. Other parts of the rat hypothalamus have been examined in a similar fashion, and the results of these studies indicate that different hypothalamic areas may differ greatly in their sensitivity to the effects of alpha and beta stimulation (Leibowitz, 1970c, 1971c, and unpublished work). In general, it appears that the ventromedial hypothalamus is considerably more sensitive than the lateral hypothalamus to alpha-receptor stimulation (which elicits feeding and suppresses drinking), whereas the lateral hypothalamus is considerably more sensitive than the ventromedial hypothalamus to beta-receptor stimulation (which suppresses feeding). Furthermore, drugs that act indirectly, by releasing or depleting endogenous catecholamines or by blocking the adrenergic receptors, are found to have diametrically opposite effects on feeding when injected into the ventromedial hypothalamus versus when injected into the lateral hypothalamus. This evidence, which differentiates the medial and lateral hypothalamus via pharmacological techniques, is intriguing in light of the large number of electrical stimulation and lesion studies that also show differential effects of these two areas on feeding.

Central effects of amphetamine

Amphetamine has been known for some time to be a potent anorexic agent. Its mechanism of action, however, is still not understood. One possibility appears to be the direct or indirect stimulation by amphetamine of the beta-adrenergic receptors, possibly in the lateral hypothalamus, which inhibit feeding. This possibility was suggested by Leibowitz (1970b, 1970c), who demonstrated that the suppression of feeding induced by perifornical or lateral hypothalamic injection of amphetamine could be abolished by beta-receptor blockers and enhanced by an alpha-receptor blocker. When injected into the ventromedial hypothalamus, an area found to be very sensitive to alpha-adrenergic stimulation but relatively insensitive to beta-adrenergic stimulation (see above), amphetamine did not cause a suppression of feeding but instead caused a small enhancement of feeding.

In addition to affecting feeding behavior, amphetamine also appears to influence drinking behavior. Perifornical hypothalamic injection of amphetamine is found to suppress water intake reliably in thirsty rats, and this effect, in contrast to amphetamine's potent beta-suppressing effect on food intake, is found to be mediated by alpha receptors (Leibowitz, 1973). These findings indicate that in the brain, as in the periphery, amphetamine can have both alpha- and beta-receptor action.

Central effects of chlorpromazine

Chlorpromazine, a widely used tranquilizing agent, is generally thought to block alpha-adrenergic receptors, at least those in the peripheral nervous system. In order to test whether this drug has such action on the hypothalamic alpha receptors that elicit feeding, Leibowitz and Miller (1969) injected chlorpromazine directly into the perifornical hypothalamus of the rat. Instead of suppressing feeding as do other alpha-receptor blockers, chlorpromazine was unexpectedly found to stimulate feeding reliably. This effect, like that induced by NE, could be clocked by an alpha blocker but not by a beta blocker (Leibowitz, 1972b, in preparation). It could also be reliably suppressed by drugs that deplete endogenous stores of NE but that leave intact postsynaptic receptors that are sensitive to NE (Leibowitz, 1969). This evidence suggests that, at least in the perifornical hypothalamus, chlorpromazine *activates* rather than blocks the alpha-adrenergic receptor system. This finding offers a possible explanation for two intriguing clinical observations: That patients receiving chlorpromazine frequently gain weight, and that in some patients this agent has pronounced antidepressant effects.

Central effects of dopamine

The above experiments with NE demonstrate that this agonist, when injected into the rat hypothalamus, can have a variety of effects on ingestive behavior. It can elicit a drinking response as well as a feeding response, and it can suppress these two behaviors in subjects made thirsty or hungry by deprivation. Since NE appears to be synthesized and stored in the hypothalamus, it is very possible that endogenous NE may act as a neurotransmitter in systems of the brain which control feeding and drinking behavior. Epinephrine also has profound effects on ingestive responses and appears to be even more potent than NE in altering such behavior. However, whether this catecholamine is synthesized in the brain and can act as a neurotransmitter is still uncertain.

There is a good deal of evidence that dopamine, another catecholamine, is indeed synthesized and stored in the brain and that it may act as a neurotransmitter. A few investigators have examined the effects of hypothalamic injection of dopamine on feeding and drinking behavior in order to determine whether the above phenomena induced by adrenergic stimulation are specific to an adrenergic system or whether they are also mediated by a central dopaminergic system. In general, the results of these studies with dopamine reveal few effects of this catecholamine on ingestive behavior.

Slangen and Miller (1969) demonstrated in satiated rats a very weak, delayed feeding response that could very likely be a result of dopamine's being used in the synthesis of NE. Preliminary studies by Leibowitz (unpublished work) have shown that injections of dopamine, into several different hypothalamic areas and at a wide range of doses, fail to alter reliably either feeding or drinking behavior in hungry or thirsty rats. So far, therefore, the evidence does not provide support for the suggestion that a hypothalamic dopaminergic system, in addition to a hypothalamic adrenergic system, mediates the control of ingestive behavior in the rat.

Concluding remarks

The adrenergic system in the rat brain, and possibly also in the brain of the monkey, appears to play an important role in the control of ingestive behavior. Furthermore, there is evidence to suggest that the different types of adrenergic receptors, alpha and beta, which are found to exist in the periphery, also exist in the brain and that they have, at least in some cases, antagonistic effects on a particular ingestive response. The reciprocal influence of alpha receptors on feeding and drinking (stimulation and suppression, respectively) and the reciprocal influence of beta receptors on feeding and drinking (suppression and stimulation, respectively) suggest a possible central neurochemical basis for the rat's ability to balance its ingestion of food and water, at least on a short-term basis. Changes in intensity of endogenous alpha- and beta-receptor activity, and the interaction between the effects of such activity, may help the rat to exhibit one ingestive response without interference from the other ingestive response, to shift more readily from one response to the other, and to keep the magnitude of one response in balance with the magnitude of the other response.

In addition to this adrenergic receptor mechanism, there appear to be other neurochemical systems of the brain that modulate ingestive behavior. These include the cholinergic (Grossman, 1962a), the serotonergic (Goldman et al., 1971), and the angiotensin (Epstein et al., 1970) systems. It seems very possible that these systems and the adrenergic system interact in their regulation of feeding and drinking. If this is indeed the case, then a comprehensive and accurate model of the central control of ingestive behavior will have to be founded on a more complete understanding of the roles of each of the neurochemical systems, as well as of their interactions.

ACKNOWLEDGMENTS   The author's research was supported by U.S. Public Health Service grant MH 13189 and by funds from Hoffmann-La Roche and from Smith, Kline, and French.

REFERENCES

AHLQUIST, R. P., 1948. A study of adrenotropic receptors. *Amer. J. Physiol.* 153:586–600.

BOOTH, D. A., 1967. Localization of the adrenergic feeding system in the rat diencephalon. *Science* 158:515–517.

BOOTH, D. A., 1968. Mechanism of action of norepinephrine in eliciting an eating response on injection into the rat hypothalamus. *J. Pharmacol. Exp. Ther.* 160:336–348.

CONTE, M., D. LEHR, W. GOLDMAN, and M. KRUKOWSKI, 1968. Inhibition of food intake by beta-adrenergic stimulation. *Pharmacologist* 10:180.

COURY, J. N., 1967. Neural correlates of food and water intake in the rat. *Science* 156:1763–1765.

EPSTEIN, A. N., J. T. FITZSIMONS, and B. J. ROLLS (née Simons), 1970. Drinking induced by injection of angiotensin into the brain of the rat. *J. Physiol. (Lond.)* 210:457–474.

FITZSIMONS, J. T., and J. LE MAGNEN, 1969. Eating as a regulatory control of drinking in the rat. *J. Comp. Physiol. Psychol.* 67:273–283.

GOLDMAN, H. W., D. LEHR, and E. FRIEDMAN, 1971. Antagonistic effects of alpha and beta-adrenergically coded hypothalamic neurons on consummatory behavior in the rat. *Nature (Lond.)* 231:453–455.

GROSSMAN, S. P., 1962a. Direct adrenergic and cholinergic stimulation of hypothalamic mechanisms. *Amer. J. Physiol.* 202:872–882.

GROSSMAN, S. P., 1962b. Effects of adrenergic and cholinergic blocking agents on hypothalamic mechanisms. *Amer. J. Physiol.* 202:1230–1236.

HOUPT, K. A., and A. N. EPSTEIN, 1971. The complete dependence of beta-adrenergic drinking on renal dipsogen. *Physiol. Behav.* 7:897–902.

INNES, I. R., and M. NICKERSON, 1970. Drugs acting on postganglionic adrenergic nerve endings and structures innervated by them (sympathomimetic drugs). In *The Pharmacological Basis of Therapeutics,* L. S. Goodman and A. Gilman, eds. New York: Macmillan, 4th Ed., pp. 478–523.

KISSILEFF, H. R., 1969. Food-associated drinking in the rat. *J. Comp. Physiol. Psychol.* 67:284–300.

LEHR, D., J. MALLOW, and M. KRUKOWSKI, 1967. Copious drinking and simultaneous inhibition of urine flow elicited by beta-adrenergic stimulation and contrary effect of alpha-adrenergic stimulation. *J. Pharmacol. Exp. Ther.* 158:150–163.

LEIBOWITZ, S. F., 1969. Mechanism of unexpected adrenergic effect from hypothalamic injection of chlorpromazine. *Proc. 77th Annual Convention, APA,* pp. 901–902.

LEIBOWITZ, S. F., 1970a. Hypothalamic beta-adrenergic "satiety" system antagonizes an alpha-adrenergic "hunger" system in the rat. *Nature (Lond.)* 226:963–964.

LEIBOWITZ, S. F., 1970b. Amphetamine's anorexic versus hunger-inducing effects mediated respectively by hypothalamic beta- versus alpha-adrenergic receptors. *Proc. 78th Annual Convention, APA,* pp. 813–814.

LEIBOWITZ, S. F., 1970c. Reciprocal hunger-regulating circuits involving alpha- and beta-adrenergic receptors located, respectively, in the ventromedial and lateral hypothalamus. *Proc. Nat. Acad. Sci. USA* 67:1063–1070.

LEIBOWITZ, S. F., 1971a. Hypothalamic alpha- and beta-adrenergic systems regulate both thirst and hunger in the rat. *Proc. Nat. Acad. Sci. USA* 68:332–334.

LEIBOWITZ, S. F., 1971b. Hypothalamic β-adrenergic "thirst" system mediates drinking induced by carbachol and transiently by norepinephrine. *Fed. Proc.* 30:280.

LEIBOWITZ, S. F., 1971c. Hypothalamic norepinephrine as an alpha- and beta-adrenergic neurotransmitter active in the regulation of normal hunger. *Proc. 78th Annual Convention, APA* p. 741.

LEIBOWITZ, S. F., 1972a. Hypothalamic beta-adrenergic receptors and their interaction with alpha receptors in the regulation of drinking. Paper presented at 43rd Annual Meeting, EPA, Boston, April 27–29.

LEIBOWITZ, S. F., 1972b. Hypothalamic alpha-adrenergic suppression of drinking: Effects on several types of thirst. *Proc. 80th Annual Convention, APA*, 845–846.

LEIBOWITZ, S. F., 1972c. Central adrenergic receptors and the regulation of hunger and thirst. In *Neurotransmitters*. Res. Publ. A.R.N.M.D., Vol. 50, 1972, pp. 327–358.

LEIBOWITZ, S. F., 1973. Alpha-adrenergic receptors mediate suppression of drinking induced by hypothalamic amphetamine injection. *Fed. Proc.* 32:754.

LEIBOWITZ, S. F., and N. E. MILLER, 1969. Unexpected adrenergic effect of chlorpromazine: Eating elicited by injection into rat hypothalamus. *Science* 165:609–611.

MILLER, N. E., 1965. Chemical coding of behavior in the brain. *Science* 148:328–338.

MILLER, N. E., K. S. GOTTESMAN, and N. EMERY, 1964. Dose response to carbachol and norepinephrine in rat hypothalamus. *Amer. J. Physiol.* 206:1384–1388.

MYERS, R. D., 1969. Chemical mechanisms in the hypothalamus mediating eating and drinking in the monkey. *Ann. N.Y. Acad. Sci.* Art. 2, 157:918–933.

MYERS, R. D., and T. L. YAKSH, 1968. Feeding and temperature responses in the unrestrained rat after injection of cholinergic and aminergic substances into the cerebral ventricles. *Physiol. Behav.* 3:917–928.

RUSSEK, M., J. A. F. STEVENSON, and G. J. MOGENSON, 1968. Anorexigenic effects of adrenaline, amphetamine, and FMSIA. *Can. J. Physiol. Pharmacol.* 46:635–638.

SINGER, G., and J. KELLY, 1972. Cholinergic and adrenergic interaction in the hypothalamic control of drinking and eating behavior. *Physiol. Behav.* 8:885–890.

SLANGEN, J. L., and N. E. MILLER, 1969. Pharmacological tests for the function of hypothalamic norepinephrine in eating behavior. *Physiol. Behav.* 4:543–552.

63 Catecholamines as Mediators of Drug Effects in Schizophrenia

SOLOMON H. SNYDER

ABSTRACT Interactions of psychoactive drugs and neuro-transmitters are a fruitful area for correlating biochemistry and behavior, especially as related to psychiatric illness. Ways in which psychoactive drugs influence neurotransmitters, particularly the biogenic amines, are reviewed. Emphasis is placed on a critical assessment of criteria for determining if a given "effect" of a drug represents its mode of action. Possible clues that such drug effects afford to the pathophysiology of various psychiatric disabilities in schizophrenia are evaluated.

AN ULTIMATE GOAL of the topic *Biochemistry and Behavior* might be to link specific human behaviors to particular neurochemical events. For many people, the highest priority lies with the aberrant behavior of the mentally ill. And of the major mental illnesses, the one with the most profound financial, human, and scientific consequences is schizophrenia.

Neurotransmitter disposition represents the biochemistry that is most unique to nervous tissues. Accordingly, the greatest success in ascribing specific behavioral events to individual biochemical sequences has arisen from studies of neurotransmitters, especially the biogenic amines serotonin and the catecholamines, norepinephrine, and dopamine. The important histochemical demonstration of monoamine tracts by fluorescence histochemistry (Hillarp et al., 1966; Ungerstedt, 1971) has greatly facilitated the tasks of workers such as Ungerstedt and S. Iversen (this volume), who have used stereotaxically administered 6-hydroxydopamine to produce selective lesions of particular catecholamine tracts in the brain with 6-hydroxydopamine. In this way they have been able to show that particular behavioral effects of drugs can be ascribed to one or another of the various catecholamine pathways in the brain. Unfortunately, no such direct techniques are available for exploring the neurochemical substrata of altered behavior in man. Moreover, in the case of the disturbed behavior of psychiatric patients, there can be no adequate animal models. Despite these difficulties, we shall endeavor in this essay to relate clinical features of schizophrenia, the actions of anti-

schizophrenic and psychotogenic drugs in man and animals, and present understanding of catecholamine disposition. The aim is to evaluate critically a possible role of catecholamines in mediating drug effects and, conceivably, pathophysiologic mechanisms in schizophrenia.

What makes biochemical theorizing about schizophrenia particularly difficult is our uncertainty as to whether the disease is a single entity or a cluster of illnesses. Lacking an obvious organic pathology, psychiatrists are forced to make diagnoses by relying upon certain arbitrarily selected clinical symptoms. Accordingly it is not surprising that from country to country and culture to culture there are wide disparities in the criteria for a diagnosis of schizophrenia. Despite these difficulties, psychiatrists who have carefully analyzed specific symptom complexes and genetic factors now are in agreement that "classic" schizophrenia, embodying patients about whose diagnosis most psychiatrists would concur, is determined by a prominent genetic component. This and other forms of the disease that are genetically related have certain clinical features in common. The psychological disturbances most characteristic of patients with classic schizophrenia are those defined by Bleuler (1911) as the "fundamental" symptoms of schizophrenia. These include a peculiar thought disorder, a disturbance of feeling or "affective" responses to the environment, and "autism," a withdrawal from meaningful interactions with other people. Bleuler felt that hallucinations and delusions, which are certainly among the most dramatic manifestations of schizophrenia, are only secondary symptoms, because they are not constant or essential to the disease. Although lack of space dictates that we present the symptoms here in an overly simplified fashion, the notion of focusing upon particular behaviors as being either primary or secondary symptoms of schizophrenia is important in seeking biochemical correlates. For instance, if a particular drug were to regularly evoke hallucinations but no other symptoms of schizophrenia, we would doubt that it is of value in explicating the fundamental disturbance in the brains of schizophrenic subjects.

In probing the biochemical substrata of the schizophrenic disturbance, drugs have been useful in two ways:

SOLOMON H. SNYDER Department of Pharmacology and Psychiatry, Johns Hopkins Medical School, Baltimore, Maryland

Phenothiazine drugs greatly alleviate the symptoms of schizophrenia. If they truly are "antischizophrenic," in that they act primarily on the fundamental schizophrenic process, then understanding their mechanism of action might conceivably shed light upon abnormal brain functioning in schizophrenia. Another approach utilizes drugs that elicit "model psychoses." Should we conclude that certain drug psychoses provide a meaningful model of schizophrenia, then knowing their neurochemical bases would also help elucidate the pathophysiology of schizophrenia.

Are phenothiazines antischizophrenic?

The phenothiazines and related drugs have revolutionized the treatment of schizophrenia, enabling many patients to function normally, or almost normally, in society. There has been much debate as to whether they act upon something fundamental to schizophrenia or whether they merely are some sort of supersedatives. One way of examining this question would be to compare the clinical efficacy of phenothiazines and sedatives. In large-scale, double-blind, and well-controlled multihospital collaborative studies sponsored by the National Institutes of Mental Health (NIMH) and the Veterans Administration (VA), a variety of phenothiazines have been compared to sedatives, especially phenobarbital (Figure 1) (Davis, 1965). Phenobarbital was no better than placebo in any of these studies, whereas most phenothiazine drugs were significantly more effective than phenobarbital and placebo. One might ask whether these are fair comparisons. Perhaps more powerful antianxiety sedatives, such as diazepam (Valium) or chlordiazepoxide (Librium), might prove more efficacious than phenobarbital and compete with the phenothiazine drugs in the treatment of schizophrenia. However, fairly extensive trials of these agents have also shown them to be ineffective in the treatment of schizophrenia, despite their accepted efficacy in relieving anxiety. Indeed, because drugs such as diazepam and chlordiazepoxide are more effective than the phenothiazines in relieving anxiety, one can conclude that anxiety per se is not a fundamental characteristic of schizophrenia.

The NIMH-VA collaborative studies also provided another means of judging the extent to which phenothiazines exert a selectively antischizophrenic action. Since large numbers of patients were rated for a variety of symptoms, one could analyze the extent to which particular clinical features of the disease responded to the drugs (Table I) (Klein and Davis, 1969). What Bleuler refers to as the fundamental symptoms of schizophrenia tended to show the greatest response to drug treatment.

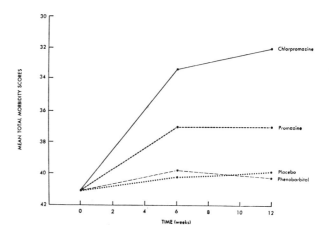

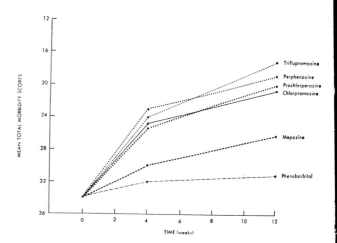

FIGURE 1   Comparative therapeutic efficacy of various phenothiazines over time in the treatment of schizophrenia. Data are derived from collaborative NIMH-VA studies and adapted, from Klein and Davis (1969).

TABLE I

Differential response of schizophrenic symptoms to phenothiazines

Bleuler's Classification	Response
Fundamental Symptoms:	
Thought disorder	+ + +
Blunted affect	+ + +
Withdrawal	+ + +
Autistic behavior	+ + +
Accessory Symptoms:	
Hallucinations	+ +
Paranoid ideation	+
Grandiosity	+
Hostility-belligerence	0
Nonschizophrenic Symptoms:	
Anxiety-tension-agitation	0
Guilt-depression	0

Adapted from Klein and Davis (1969).

Secondary symptoms, such as delusions and hallucinations, responded somewhat less, and nonschizophrenic symptoms, such as anxiety and depression, failed to show any specific improvement with phenothiazines. By contrast, sedatives would relieve agitation with much less influence upon thought disorder or the abnormality of affective response to the environment.

Thus one can make a reasonable case that the phenothiazines exert a unique therapeutic action in schizophrenic patients. We should be cautious before making the next conceptual leap to the conclusion that the drugs exert a biochemical effect that directly reverses whatever is biochemically abnormal in the brains of schizophrenics. The phenothiazines might act at a secondary site not directly related to the schizophrenic abnormality but through which the drugs can ease the distress of patients so that the schizophrenic symptoms can resolve themselves. This would be analogous to applying a tourniquet upstream from a site of bleeding, slowing the gush of blood from a wound so that the body's clotting mechanisms can then relieve the primary abnormality, a tear in the wall of the blood vessel. In favor of such a not-so-direct antischizophrenic action of phenothiazines is the well-known fact that these drugs do not "cure" schizophrenia but only facilitate remissions. Indeed, failure to maintain schizophrenic patients on phenothiazines while they are in remission results in a much greater incidence of relapse (Klein and Davis, 1969). Despite this caveat, we feel that it is worthwhile to explore the mechanism of action of the phenothiazines, because there is a reasonable possibility that the neurochemical effects of these drugs might have direct bearing on brain dysfunction in schizophrenia.

Phenothiazine pharmacology

Phenothiazines are highly reactive chemicals capable of pi electron donation or acceptance, hydrophobic binding, and ionic links via the side-chain amine. Accordingly these drugs exert biochemical effects upon almost every system that has been examined (Guth and Spirtes, 1963). How might one decide which of these actions is most relevant to the therapeutic efficacy of the drugs? Of the large numbers of phenothiazines that have been employed clinically and that are fairly similar in their chemical structure, some are highly effective in the treatment of schizophrenia; others are somewhat less effective clinically, while yet others are definitely ineffective. Biochemical effects that would correlate with known clinical actions would be the best candidates to mediate the therapeutic effects of the drugs.

To my knowledge the best correlation with clinical potency has been evinced by certain actions of phenothiazines and related drugs upon brain catecholamines, especially dopamine. Carlsson and Lindqvist (1963) reported that chlorpromazine and other antischizophrenic drugs significantly elevated brain levels of the methoxylated metabolites of norepinephrine and dopamine, while the antihistaminic phenothiazine, promethazine, was ineffective. Haloperidol, the butyrophenone whose antischizophrenic effects are elicited at much lower doses than those of chlorpromazine, was correspondingly more potent in elevating the methoxylated metabolites of the catecholamines. On the basis of these limited findings, Carlsson and Lindqvist speculated that the phenothiazines block catecholamine receptor sites and that, via a neuronal feedback loop, a message is conveyed to the catecholamine cells, "we receptors aren't getting enough transmitter; send us more catecholamines!" Accordingly, the catecholamine neurons proceed to fire more rapidly and, as a corollary, they synthesize more catecholamines and release more metabolites. Carlsson's speculations have been confirmed in studies that show that phenothiazines and butyrophenones, in proportion to their clinical efficacy, accelerate catecholamine synthesis (Figure 2) (Nyback et al., 1968). The influence of these drugs upon dopamine synthesis correlates with clinical effects better than their actions on norepinephrine synthesis. Indeed, several extremely potent butyrophenone tranquilizers selectively accelerate dopamine turnover with negligible effects upon norepinephrine. By neurophysiologic techniques, Aghajanian et al. (1973) directly demonstrated that phenothiazines can speed up the firing of dopamine cells in the brain stem, and more recently Aghajanian (personal communication) has obtained evidence by iontophoretic experiments that antischizophrenic phenothiazine tranquilizers do block the responses to dopamine of cells which normally receive dopamine neuronal input. Thus, both biochemical and neurophysiological studies have demonstrated interactions of the phenothiazine tranquilizers with dopamine systems in the brain, which are consistent with a blockade of dopamine receptors and which correlate well with their antischizophrenic actions.

Part of the difficulty in securing general acceptance to the hypothesis of a relationship between dopamine receptor blockade and the clinical actions of phenothiazines is the lack of chemical similarity between phenothiazines and catecholamines. Phenothiazines are complex multiringed structures, while dopamine is a simple phenylethylamine. Recently Horn and Snyder (1971) noted that chlorpromazine, in its preferred conformation as determined by X-ray crystallography, can be superimposed, in part, upon the accepted preferred

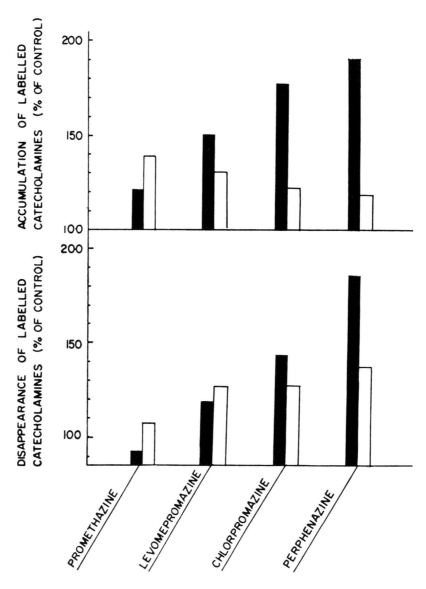

FIGURE 2 The effect of four phenothiazine derivatives on the turnover of dopamine (filled columns) and norepinephrine (open columns) in mouse brain. Mean effects of 0.5, 2, and 10 mg/kg of the drugs are shown. Adapted from Nyback (1971).

conformation of dopamine or norepinephrine, providing a molecular mechanism whereby phenothiazines might block dopamine receptors. In the preferred conformation of chlorpromazine, its side chain tilts away from the midline toward the chlorine-substituted ring (Figure 3). Presumably the chlorine on Ring A is responsible for the "tilt" of the side chain, because if there were no substituent on Ring A, both rings A and C would be symmetrical, and one would expect the side chain to be fully extended. Accordingly, phenothiazines lacking a substituent on Ring A should mimic the conformation of dopamine less effectively and thus have less affinity for dopamine receptors and presumably be less efficacious in the treatment of schizophrenia. Of the dozen or so phenothiazine tranquilizers that have been widely employed clinically, only two lack a substituent on Ring A. Strikingly, mepazine and promazine, the two phenothiazines lacking a Ring A substituent, are significantly less effective as antischizophrenic drugs than the others. (Table II) (Cole and Davis, 1969).

Besides the Ring A substituent, another major requirement for therapeutic activity of phenothiazines is separation of the side-chain amine by three carbon atoms from the ring system. Molecular models indicate that shortening the side chain to two carbon atoms would make the assumption of the dopamine-like conformation less likely. Just as predicted, phenothiazines with two-carbon side chains, such as the antihistamine promethazine and the

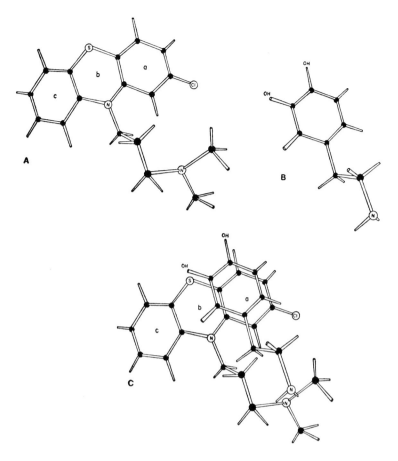

FIGURE 3 Molecular models of the preferred conformations (as determined by X-ray crystallography) of [A] chlorpromazine [B] dopamine and [C] their superimposition. Adapted from Horn and Snyder (1971).

TABLE II

Comparative efficacy of phenothiazine drugs in the treatment of schizophrenia

Drug		Number of Studies in which		
Generic Name	Trade Name	Drug was more Effective than Chlorpromazine	Drug was equal to Chlorpromazine	Chlorpromazine was more Effective
Mepazine	Pacatal	0	0	4
Promazine	Sparine	0	2	4
Triflupromazine	Vesprin	0	10	0
Perphenazine	Trilafon	0	6	0
Prochlorperazine	Compazine	0	10	0
Trifluoperazine	Stelazine	0	8	0
Thioridazine	Mellaril	0	12	0
Fluphenazine	Prolixin	0	7	0
Acetophenazine	Tindal	0	1	0
Thiopropazate	Dartal	0	1	0
Phenobarbital	—	0	0	6

Data are adapted from Klein and Davis (1969) and are drawn from a review of double-blind, well-controlled clinical studies.

antiparkinsonian agent diethazine, lack therapeutic effectiveness in schizophrenia.

Besides being associated with their antischizophrenic activity, dopamine receptor blockade by phenothiazines and butyrophenones may explain the prominent extrapyramidal side effects of these drugs. The symptoms of Parkinson's disease are now thought to be associated with a deficiency of dopamine in the caudate putamen, the area of major dopamine terminals in the brain, and the therapeutic action of L-dopa is presumed to arise from an alleviation of this dopamine deficiency. What dopamine pathways in the brain might be related to the antischizophrenic actions of phenothiazines? Besides the nigrostriatal dopamine tract, other major areas with dopamine terminals include the olfactory tubercle and nucleus accumbens. (Ungerstedt, 1971.)

Amphetamine psychosis as a "model schizophrenia"

Model psychoses have provided an attractive approach to understanding schizophrenia, with the hope that their biochemical elucidation may bear upon similar defects in the brains of schizophrenics. Early interest centered upon psychedelic drug psychosis. Like schizophrenia, the mental aberrations following the ingestion of psychedelic drugs occur while the subject is alert and fully oriented to time, place, and person. This is in contradistinction to most drug psychoses that are associated with confusion, disorientation, and delirium. However, the behavioral effects of psychedelic drugs differ in important ways from schizophrenic disturbances. The major symptoms following psychedelic drug ingestion are disorders of perception. LSD-induced hallucinations are usually visual, whereas schizophrenic hallucinations tend to be auditory. Moreover there is little evidence that a schizophrenia-like thought disorder or disturbance of affect takes place in psychedelic drug induced psychosis. Hollister (1962) tabulated the symptoms of subjects receiving psychedelic drugs and found them to be markedly dissimilar to schizophrenic symptoms. Moreover he showed that tape-recorded interviews with newly admitted schizophrenics could be readily distinguished by "blind" raters from those of subjects under the influence of psychedelic drugs. In addition schizophrenics receiving psychedelic drugs report that the drug experience is quite unlike their endogenous psychosis (Chloden et al., 1955).

By contrast, amphetamine psychosis may be a better candidate for a model schizophrenia. Many cases of amphetamine psychosis have been misdiagnosed as acute paranoid schizophrenia until the history of drug use was obtained (Connell, 1958; Bell, 1965; Beamish and Kiloh, 1960), and accordingly amphetamine psychosis was suggested as an heuristic "model" schizophrenia (Kety,

1959, 1972). Amphetamine psychosis is most frequently observed in amphetamine addicts who have consumed enormous amounts of the drug over prolonged periods, e.g. 500 to 1000 mg of d-amphetamine per day for a week or more (Ellinwood, 1967). These patients develop an acute paranoid psychosis that typically resolves within a few days after termination of the drug. Unlike LSD psychosis, patients with amphetamine psychosis frequently experience auditory hallucinations very much like typical schizophrenic auditory hallucinations, including hearing vague noises and voices and occasionally having conversations with the voices. The visual hallucinations in amphetamine psychotics tend to resemble those observed in very acute schizophrenics (Bowers and Freedman, 1966; Chapman, 1966).

It should be borne in mind that more than one type of amphetamine psychosis is possible. Amphetamines can evoke an acute toxic psychosis with delirium, confusion, and disorientation as occurs with many other drugs and unlike the schizophrenia-like amphetamine psychosis. "Toxic" amphetamine psychosis occurs usually after only one or two extremely large doses rather than after chronic use of the drug. Among 42 cases of amphetamine psychosis, visual hallucinations occurred primarily among patients who became acutely psychotic, often after a single large dose, hence presumably were suffering from a "toxic" psychosis. By contrast hallucinations were usually auditory in patients whose illness developed gradually after frequent repeated doses (Connell, 1958). The "toxic" amphetamine psychoses probably account for the greater proportion of visual hallucinations in amphetamine intoxication than in schizophrenia. Here we will focus solely on the "nontoxic" amphetamine psychosis with intact sensorium and which most closely resembles clinical schizophrenia.

Some authors have criticized amphetamine psychosis as a "model schizophrenia," arguing that it might be simply related to sleep deprivation, overexcitement, or precipitation of psychosis in borderline schizophrenics. To examine these questions, Griffith et al. (1970, 1972) and Angrist and Gershon (1970) developed an experimental model of amphetamine psychosis in man by administering progressively increasing doses of d-amphetamine to volunteer subjects. None of the subjects had any evidence of preexisting schizophrenia or schizoid tendencies. Yet virtually all subjects became floridly psychotic after 1 to 4 days, indicating that amphetamine psychosis is not simply a precipitation of latent schizophrenia. Since some patients became psychotic in about 24 hr, there could not have been sufficient sleep deprivation to account for the psychosis. As for the question of overexcitement, after some initial moderate euphoria, most subjects were more sullen than excited, although it is conceivable that "internal" hyperexcitement was taking place which was not evident to the observers.

What about the fundamental schizophrenic symptoms: thought disorder, disturbance of affect, and autism? Griffith et al. (1972) did not feel that a typically schizophrenic thought disorder or affect disturbance was present in their subjects, while Angrist and Gershon (1970) did observe flattened affect and schizophrenic thinking patterns in their patients. Part of these discrepancies may stem from difficulty in interpreting the presence of abnormal affect and thinking in acutely paranoid individuals. Paranoid patients characteristically display seemingly systematic and well-organized thought patterns, although the underlying delusions are inherently disorganized. Moreover, very acute paranoid schizophrenics quite frequently display a lively affect (Cameron, 1959), while their typical thought disorder and affective disturbance only become apparent as the acute paranoid features subside.

If one takes the position that amphetamine psychosis lacks a typically schizophrenic thought disorder, then according to standard American nomenclature a better diagnostic classification would be "paranoid state," which is defined as a psychosis characterized by paranoid delusions but in which "emotional responses and behavior are consistent with the ideas held . . . and it does not manifest the bizarre fragmentation and deterioration of schizophrenic reactions" (American Psychiatric Association, 1952). However, many European psychiatrists do not feel that the entity of paranoid state exists separate from schizophrenia and would label such patients as suffering from paranoid schizophrenia.

One important reservation about treating amphetamine psychosis as a model schizophrenia is that it rarely resembles nonparanoid schizophrenia. If amphetamine psychosis is only to be a model of paranoid schizophrenia, then it might follow that paranoid schizophrenia differs in a fundamental way from other forms of schizophrenia. Contrary to this suggestion are major items of genetic and clinical evidence: (a) schizophrenics characteristically switch from one subtype of schizophrenia to another during their clinical history. (b) Genetic studies indicate that a variety of different types of schizophrenia may "run" in families (Kallman, 1938; Gottesman and Shields, 1967). These two items suggest that if amphetamine psychosis is to be a meaningful model of schizophrenia, it should show resemblance to nonparanoid forms of the disease. Why does not amphetamine psychosis mimic other forms of schizophrenia? One might speculate that the "paranoid" character is contributed by other nonschizophrenomimetic effects of the drug. Thus, perhaps amphetamine does possess a "pure" schizophrenia-mimicking action, but some other effect of the drug transforms the clinical picture into a predominantly paranoid one. A good candidate for such an effect would be the well-known central stimulant-alerting actions of amphetamine. One might speculate that the major feature that differentiates paranoid schizophrenics from other schizophrenics is their hyperalert striving to cast the bewildering array of psychotic transformations into a coherent and meaningful process. It is this paranoid hypervigilance that, by enabling the patient to loculate his disturbance, tends to mask the typically schizophrenic thought disorder and disturbance of affect.

Another finding linking amphetamine actions to schizophrenia is the observation that amphetamine and its analogues directly exacerbate schizophrenic symptoms (Levine et al., 1948; Hope et al., 1951; Pennes, 1954; Janowsky et al., 1973). It is important to note that amphetamines intensify the schizophrenic symptoms, rather than adding different psychotic symptoms to the schizophrenic illness. Patients themselves perceive that their illness is worsening under the influence of the drug. By contrast, when schizophrenics are treated with LSD, they can recognize that the psychedelic drug psychosis is something different from their own mental disturbance. Amphetamines fail to elicit psychotic effects (in the small doses employed) in schizophrenics in remission or in manic or depressed patients. (Janowsky et al., 1973.)

Another item favoring an association between amphetamine-induced mental disturbance and schizophrenia is the fact that phenothiazines and butyrophenones are the best antidotes for amphetamine psychosis (Angrist, Griffith, and Davis, personal communications). This is especially striking, because barbiturate sedatives fail to alleviate amphetamine psychosis and, in some cases, accentuate the symptoms (Angrist, personal communication).

Of other drug psychoses, those elicited by cocaine and L-dopa are of particular interest. Cocaine, whose central effects are thought to be related to potentiation of brain catecholamines, produces a psychosis whose clinical features are essentially the same as those of amphetamine psychosis (Mayer-Gross et al., 1960). L-dopa is the precursor of the catecholamines dopamine and norepinephrine. In Parkinsonian patients treated with L-dopa, about 15 to 30% display psychiatric side effects (McDowell, 1970). Symptoms resembling amphetamine psychosis are rare in these patients (Celesia and Barr, 1970). The most common psychiatric reaction is a delirious toxic psychosis with confusion and visual hallucinations. Since dopa certainly does act via brain catecholamines, why does it only rarely elicit a typical amphetamine psychosis? The answer probably lies in the relative doses employed. Amphetamine psychosis is rarely observed unless subjects are ingesting well in excess of 100 mg of *d*-amphetamine per day. Although direct comparisons are difficult, it is probable that to secure comparable central effects with L-dopa would require about 20 gm of the drug, whereas the typical dose in Parkinsonian patients is only about 4 gm. Moreover, Parkinsonian patients are generally more than 50 years old, hence would be much more susceptible to organic brain symptoms than schizophrenia-like symptoms.

Because of these difficulties, a more efficient means of evaluating the influence of L-dopa upon schizophrenia-like processes would be to ascertain whether it might accentuate schizophrenic symptoms. Yaryura-Tobias et al. (1970) and B. Angrist (personal communication) have observed that L-dopa does indeed activate the symptoms of schizophrenia in some patients. One must be cautious in interpreting these results, however, because there is evidence that L-dopa may exert a generalized "activating" effect in several types of psychiatric patients (Goodwin et al., 1970).

Amphetamine-catecholamine interactions

How does amphetamine act in the brain? Because of obvious chemical similarities (Figure 4), amphetamine is generally thought to act via the brain catecholamines dopamine and norepinephrine. The most efficient means of exploring exactly which tract mediates behavioral effects of amphetamines in animals is to make discrete lesions with 6-hydroxydopamine. 6-Hydroxydopamine is accumulated by catecholamine neurons selectively, after which it auto-oxidizes and thereby destroys the neurons (Thoenen, 1972; Ungerstedt, 1973a). Destruction of the nigrostriatal dopamine pathway by implanting 6-hydroxydopamine directly into the substantia nigra abolishes the stereotyped compulsive gnawing, licking, and sniffing elicited by amphetamine (S. Iversen, this volume; Ungerstedt, 1973b). S. Iversen (this volume) reported that these same lesions failed to alter the locomotor stimulation normally evoked by amphetamine, indicating that this effect is not a function of the nigrostriatal tract.

To evaluate the differential roles of dopamine and norepinephrine in mediating amphetamine actions in man, we have employed isomers of amphetamine. We took advantage of the differences in stereoselectivity of dopamine and norepinephrine neurons (Coyle and Snyder, 1969; Hendley and Snyder, 1972; Hendley et al., 1972; Snyder et al., 1970a, 1970b; Snyder, 1970). (Table III.) Norepinephrine neurons exhibit about four times as much affinity for the naturally occurring *l*-form of norepinephrine (Coyle and Snyder, 1969; Iversen et al., 1971) as for *d*-norepinephrine. By contrast, dopamine neurons in the corpus striatum (Coyle and Snyder, 1969; Iversen et al., 1971) and retina (Hendley and Snyder, 1972) fail to

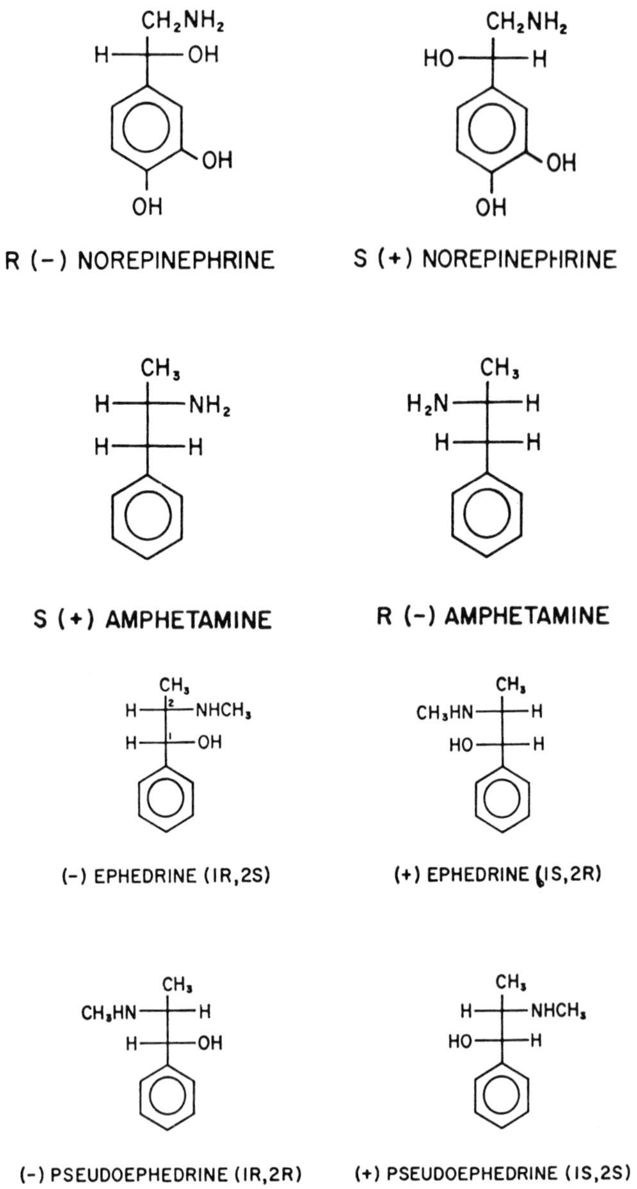

FIGURE 4 The structures and absolute configurations of stereoisomers of norepinephrine, amphetamine, and ephedrine. Ephedrine represents the *erythro* type of configuration and pseudoephedrine, the *threo* type. Adapted from Hendley et al., 1972.

differentiate between norepinephrine isomers, presumably because dopamine, their normal catecholamine transmitter, is a symmetric molecule lacking stereoisomers.

Isomers of norepinephrine differ at the β carbon, while isomers of amphetamine differ at the α carbon. Amphetamine is a potent inhibitor of the catecholamine reuptake system, whereby neuronal accumulation of synaptically released dopamine and norepinephrine presumably terminates their synaptic activities (Iversen, 1967; Axelrod,

TABLE III

Affinity of norepinephrine, amphetamine, and ephedrine isomers for the catecholamine uptake mechanism of brain synaptosomes

Compound	Absolute Config- uration	Relative Affinity	
		Dopamine Neurons (corpus striatum)	Norepine- phrine Neurons (cerebral cortex)
(−)-Norepinephrine	1-R	40.0	240.0
(+)-Norepinephrine	1-S	40.0	60.0
(+)-Amphetamine	2-S	720.0	240.0
(−)-Amphetamine	2-R	720.0	24.0
(−)-Ephedrine	1R:2S	11.6	100.0
(+)-Ephedrine	1S:2R	6.0	25.7
(+)-Pseudoephedrine	1S:2S	3.5	3.4
(−)-Pseudoephedrine	1R:2R	1.7	1.0

Affinities for ephedrine isomers are derived from experiments in which the dose required for 50% inhibition of the uptake of 0.1 μm ($\pm$)-^{3}H-norepinephrine (ID-50) was assessed (Hendley et al., 1972). Affinity is actually the relative potency derived from the ID-50, thusly: Relative potency = (7.2 × 10^{-7}M)/ ID-50 × 100. Affinities for amphetamine and norepinephrine isomers are obtained from Ki values for the reduction of ($\pm$)-^{3}H-norepinephrine uptake (Coyle and Snyder, 1969), thusly: Affinity = 7.2 × 10^{-7}M/Ki × 100. All experiments employed synaptosome-rich homogenates of cerebral cortex or corpus striatum.

1965). The central stimulant *d*-amphetamine is considerably more potent in blocking catecholamine uptake by norepinephrine terminals than is *l*-amphetamine, a much weaker central stimulant; while in dopamine neurons the amphetamine isomers have similar potency (Coyle and Snyder, 1969; Hendley and Snyder, 1972). With drugs such as the ephedrines, which possess both asymmetric α and β carbons (Figure 4) these differences may be amplified (Table III). Thus while *d*- and *l*-amphetamines differ tenfold in their effects upon norepinephrine neurons, the most and least potent ephedrine isomers differ by 100-fold. With ephedrine isomers there is less than 1/10 as much stereoselectivity toward ephedrine isomers in dopamine than in norepinephrine neurons (Hendley et al., 1972). Because the stereoselectivity of the catecholamine uptake process is much less striking than the stereospecificity of receptor activation by catecholamines, great care must be exercised in preparation of tissue and design of blank values to correct for nonspecific accumulation of amine. One must also ensure maximal uptake velocity, linearity with time and tissue concentration, and optimal surface area to obtain stereoselective effects (Hendley and Snyder, 1972; Hendley et al., 1972). Such technical issues may explain difficulties in demonstrating stereoselective catecholamine uptake in some tissues (Ferris et al., 1972).

The differential effects of amphetamine isomers on catecholamine uptake by dopamine and norepinephrine neurons can be demonstrated in vivo as well as in vitro (Taylor and Snyder, 1970, 1971; Snyder et al., 1970b). Moreover, norepinephrine depletion, thought to reflect the norepinephrine releasing action of amphetamine, another major mechanism of its pharmacological activity, also responds to amphetamine isomers in a pattern similar to influences upon uptake, while methoxylated amine metabolites correlate less well (Svensson, 1971; Scheel-Kruger, 1972).

From these stereoselective actions of amphetamines emerges a simple paradigm for drawing inferences as to whether brain dopamine or norepinephrine mediates particular behavioral effects of amphetamines. Behaviors mediated by norepinephrine should be affected much more by *d*- than by *l*-amphetamine, while the two isomers should have similar potencies in eliciting dopamine mediated behaviors. *d*-Amphetamine is only about twice as potent as *l*-amphetamine in evoking stereotyped compulsive gnawing behavior in rats (Taylor and Snyder, 1970, 1971). This suggests a prominent role for dopamine in this behavior and concurs with the conclusions of several workers using other approaches (Creese and Iversen, 1972; S. Iversen, this volume; Ungerstedt, this volume; Ernst and Smelik, 1966; Fog et al., 1967). By contrast, *d*-amphetamine is ten times as potent as *l*-amphetamine in stimulating locomotor activity in rats, a finding that closely parallels the tenfold difference in the potency of these drugs in blocking catecholamine uptake by norepinephrine neurons. These results, which favor a norepinephrine mediation of this behavior, are consistent with the findings of S. Iversen (this volume) and Creese and Iversen (1972) that 6-hydroxydopamine lesions of the nigrostriatal system do not diminish amphetamine-induced locomotor stimulation.

In man, *d*-amphetamine is about 5 times as potent a central stimulant as *l*-amphetamine, suggesting that norepinephrine mediates the alerting actions of amphetamine in man (Prinzmetal and Alles, 1940). This would seem reasonable, because the alerting action of amphetamine is generally thought to be a cerebral cortical event and because the cerebral cortex is extensively innervated by norepinephrine fibers but receives no dopamine innervation.

Amphetamine is a valuable treatment modality for hyperactive children with "minimal brain dysfunction." Comparisons of *d*- and *l*-amphetamine in minimal brain dysfunction in a longitudinal study of one patient (Snyder and Meyerhoff, 1973) and in a double blind evaluation of 11 patients (Arnold et al., 1972) failed to reveal a statistically significant difference in the efficacy of the isomers, suggesting that the beneficial effect of amphetamine in minimal brain dysfunction is dopamine mediated. The facial and body tickings of Gilles de la Tourette's disease are reminiscent of motor dysfunctions that occur in Parkinsonian patients treated with L-dopa. In one patient with Gilles de la Tourette's disease studied over a 2-month period, d-amphetamine greatly accentuated the ticking while l-amphetamine was without effect (Snyder and Meyerhoff, 1973). This suggests that the ticking in Gilles de la Tourette's disease, which is most effectively ameliorated by the butyrophenone haloperidol, is related more to norepinephrine than to dopamine.

On the basis of the animal studies of amphetamine isomers, Angrist et al. (1971) deduced that comparison of the potencies of d- and l-amphetamine in eliciting amphetamine psychosis might provide an indication of the relative roles of norepinephrine and dopamine. Accordingly, on three separate occasions, three volunteers were given amphetamines in progressively increasing doses until psychosis was clearly evident. On one occasion d-amphetamine was utilized, while another time l-amphetamine was employed, and on the third occasion, a mixture of equal parts of d- and l-amphetamine was administered. Drug administration was terminated when psychosis was judged to be present, which occurred with a sudden emergence of paranoid delusions and a cold detached affect, so that one could readily estimate the total dose of drug required to elicit psychosis.

All three subjects were rendered psychotic by all three drugs. Strikingly, d- and l-amphetamine had quite similar potencies in eliciting the psychosis. In the three patients, the ratios of the total dose of l-amphetamine to the total dose of d-amphetamine required to produce the psychosis were 1.25, 1.53, and 1.0. The quantity of racemic amphetamine required to produce psychosis was intermediate between the amounts necessary for d and l forms.

During the early stages of amphetamine administration, Angrist and Gershon (1971) observed much more central stimulation with d-amphetamine than with l-amphetamine. Thus in these human subjects d- and l-amphetamine were much more similar in their capacity to elicit psychosis than in their ability to effect central stimulation.

Earlier we asked why amphetamine invariably elicited paranoid psychoses and suggested that the psychosis-eliciting and the central stimulant actions of the drug might be quite distinct. Amphetamine could provoke a schizophrenia-like psychosis, whose paranoid flavor was conferred by the central stimulant actions of the drug. The experimental results with amphetamine isomers suggest that the psychosis elicited by amphetamine may be mediated by brain dopamine, while central stimulant effects are elicited by brain norepinephrine. (Table IV.) One might speculate that the norepinephrine-mediated

alerting action forces the patient to strive for an intellectual framework in which to focus all the strange feelings that are coming over him as the psychosis develops. This quest for meaning and its subsequent "discovery" in a system of delusions might be the essence of the paranoid process in these patients. According to this reasoning, a "schizophrenia-like" component of amphetamine psychosis may be mediated primarily by brain dopamine, while the "paranoid solution" is facilitated by the drug's alerting effects via stimulation of norepinephrine systems.

TABLE IV

Relations between drugs, catecholamines, and schizophrenia

Phenothiazines (and related antischizophrenic drugs).
(1) Phenothiazines have true antischizophrenic actions:
   (a) Are more effective than sedatives.
   (b) Act best on fundamental symptoms.
(2) Blockade of dopamine receptors by phenothiazines is proportional to their clinical efficacy.
(3) The ability of phenothiazines to mimic the preferred catecholamine conformation predicts their therapeutic activity.

Amphetamines
(1) In small doses, amphetamines specifically activate schizophrenic symptoms.
(2) Amphetamines can evoke a psychosis indistinguishable from acute paranoid schizophrenia.
(3) Phenothiazines are the best antidotes for amphetamine psychosis.
(4) Isomer studies suggest that dopamine mediates amphetamine psychosis, while norepinephrine mediates the central stimulant effects of amphetamine.
(5) Speculation: Via dopamine, amphetamine may evoke a schizophrenia-like psychosis that is transformed by norepinephrine-mediated central stimulant actions of amphetamine into a paranoid psychosis.

If indeed, the norepinephrine-mediated alerting actions of amphetamines interact with and "contaminate" a purer dopamine-mediated amphetamine psychosis, then removal of these "norepinephrine effects" might leave one with an amphetamine psychosis more closely resembling schizophrenia than the present version. Conceivably, development of a drug that would stimulate dopamine but not norepinephrine mechanisms in the brain would produce such a "pure" model of schizophrenia.

Of course, as pointed out earlier in our discussion of phenothiazines, the fact that a schizophrenia-like psychosis follows upon the action of amphetamine on brain catecholamines does not "prove" that catecholamines are involved in the "psychotomimetic" locus in the brain. Moreover even if certain catecholamine neurons were responsible for amphetamine psychosis and even if the neural disturbances in amphetamine psychosis are closely similar to those occurring in schizophrenia, we still could not conclude that these neurons are the site of primary disturbance in schizophrenia. Nonetheless, the evidence (a) that selective antischizophrenic actions of drugs appear to be mediated via brain dopamine systems (b) that a schizophrenia-like amphetamine psychosis appears to involve brain dopamine, and (c) that amphetamines can accentuate schizophrenic symptoms in a fairly specific fashion, certainly favors some role for catecholamines. Accordingly, further exploration of the differential behavioral roles of particular catecholamine systems in the brain and their interactions with drugs may considerably enhance our understanding of brain mechanisms in schizophrenia.

ACKNOWLEDGMENTS   This work was supported by U.S.P.H.S. grants MH 18501, NS 07275, DA 00266 and NIMH Research Scientist Development Award MH 33128. Experimental work providing a basis for much of the thinking in this essay was conducted in collaboration with Joseph T. Coyle, Kenneth M. Taylor, Edith D. Hendley, and Alan S. Horn.

REFERENCES

AGHAJANIAN, G. K., B. S. BUNNEY, and M. J. KUHAR, 1973. Use of single unit recording in correlating transmitter turnover with impulse flow in monoamine neurons. In *New Concepts in Neurotransmitter Regulation*, A. J. Mandell, ed. New York, (in press).

AMERICAN PSYCHIATRIC ASSOCIATION, 1952. *Diagnostic and Statistical Manual of Mental Disorder*, Vol. I. Washington, D.C.: American Psychiatric Association, p. 28.

ANGRIST, B. M., and S. GERSHON, 1970. The phenomenology of experimentally-induced amphetamine psychosis. Preliminary observations. *Amer. J. Psychiat.* 126:95–107.

ANGRIST, B., and S. GERSHON, 1971. A pilot study of pathogenic mechanisms in amphetamine psychosis utilizing differential effects of *d*- and *l*-amphetamine. *Pharmakopsychiatrie Neuro-Psychopharmacologie* 4:65–75.

ANGRIST, B. M., B. SHOPSIN, and S. GERSHON, 1971. The comparative psychotomimetic effects of stereoisomers of amphetamine. *Nature (Lond.)* 234:152–154.

ARNOLD, L. E., P. H. WENDER, K. McCLOSKEY, and S. H. SNYDER, 1972. Levoamphetamine and dextroamphetamine: Comparative efficacy in the hyperkinetic syndrome; assessment by target symptoms. *Arch. Gen. Psychiat.* 27:816–822.

AXELROD, J., 1965. The metabolism, storage and release of catecholamines. *Rec. Progr. Horm. Res.* 21:597–622.

BEAMISH, P., and L. G. KILOH, 1960. Psychoses due to amphetamine consumption. *J. Ment. Sci.* 106:337–343.

BELL, D. S., 1965. Comparison and amphetamine psychosis and schizophrenia. *Brit. J. Psychiat.* 111:701–707.

BLEULER, E., 1911. *Dementia Praecox or the Group of Schizophrenias*. New York: International University Press. English trans., J. Zinkin, 1950.

BOWERS, JR., M. J., and D. X. FREEDMAN, 1966. "Psychedelic" experiences in acute psychoses. *Arch. Gen. Psychiat.* 15: 240–248.

CAMERON, N., 1959. Paranoid conditions and paranoia. In *American Handbook of Psychiatry*, S. Arieti, ed. New York: Basic Books, 1:510.

CARLSSON, A., and M. LINDQVIST, 1963. Effect of chlorpromazine or haloperidol on the formation of 3-methoxytyramine and normetanephrine in mouse brain. *Acta Pharmacol. Toxicol.* 20:140–144.

CELESIA, G. G., and A. N. BARR, 1970. Psychosis and other psychiatric manifestations of Levodopa therapy. *Arch. Neurol.* 23:193–200.

CHAPMAN, J., 1966. The early symptoms of schizophrenia. *Brit. J. Psychiat.* 112:225–251.

CHLODEN, L. W., A. KURLAND, and C. SAVAGE, 1955. Clinical reactions and tolerance to LSD in chronic schizophrenia. *J. Nerv. Ment. Dis.* 122:211–221.

COLE, J. O., and J. M. DAVIS, 1969. Antipsychotic drugs. In *The Schizophrenic Syndrome*, H. Solomon, ed. New York: Grune and Stratton, pp. 478–568.

CONNELL, P. H., 1958. *Amphetamine Psychosis*. London: Chapman and Hall.

COYLE, J. T., and S. H. SNYDER, 1969. Catecholamine uptake by synaptosomes in homogenates of rat brain: Stereospecificity in different areas. *J. Pharmacol. Exp. Ther.* 170:221–231.

CREESE, I., and S. D. IVERSEN, 1972. Amphetamine response in rat after dopamine neuron destruction. *Nature (Lond.)* 238: 247–248.

DAVIS, J. M., 1965. Efficacy of the tranquilizing and antidepressant drugs. *Arch. Gen. Psychiat.* 13:552–572.

ELLINWOOD, JR., E.H., 1967. Amphetamine psychosis: I. Description of the individuals and process. *J. Nerv. Ment. Dis.* 144:273–283.

ERNST, A. M., and P. SMELIK, 1966. Site of action of dopamine and apomorphine on compulsive gnawing behavior in rats. *Experientia.* 22:837–839.

FERRIS, R. M., F. L. M. TANG, and R. A. MAXWELL, 1972. A comparison of the capacities of isomers of amphetamine, deoxypipradol and methylphenidate to inhibit the uptake of tritiated catecholamines into rat cerebral cortex slices, synaptosomal preparations of rat cerebral cortex, hypothalamus and striatum and into adrenergic nerves of rabbit aorta. *J. Pharmacol. Exp. Ther.* 181:407–416.

FOG, R. L., A. RANDRUP, and H. PAKKENBERG, 1967. Aminergic mechanisms in corpus striatum and amphetamine-induced stereotyped behavior. *Psychopharmacologia* 11:179–183.

GOODWIN, F. K., D. L. MURPHY, H. K. H. BRODIE, and W. E. BUNNEY, JR., 1970. L-Dopa, catecholamines and behavior; a clinical and biochemical study in depressed patients. *Biol. Psychiat.* 2:341–366.

GOTTESMAN, I. I., and J. SHIELDS, 1967. A polygenic theory of schizophrenia. *Proc. Natl. Acad. Sci. USA* 58:199–205.

GRIFFITH, J. D., J. H. CAVANAUGH, J. HELD, and J. A. OATES, 1970. Experimental psychosis induced by the administration of *d*-amphetamine. In *Amphetamines and Related Compounds*. New York: Raven Press, pp. 897–904.

GRIFFITH, J. D., J. CAVANAUGH, J. HELD, and J. A. OATES, 1972. Dextroamphetamine: Evaluation of psychotomimetic properties in man. *Arch. Gen. Psychiat.* 26:97–100.

GUTH, P. S., and M. A. SPIRTES, 1963. The phenothiazine tranquilizers: Biochemical and biophysical actions. *Int. Rev. Neurobiol.* 7:231–278.

HENDLEY, E. D., and S. H. SNYDER, 1972. Stereoselectivity of catecholamine uptake in noradrenergic and dopaminergic peripheral organs. *Eur. J. Pharmacol.* 19:56–66.

HENDLEY, E. D., S. H. SNYDER, J. J. FAULEY, and J. B. LaPIDUS, 1972. Stereoselectivity of catecholamine uptake by brain synaptosomes: Studies with ephedrine, methylphenidate and phenyl-2-piperidyl carbinol. *J. Pharmacol. Exp. Ther.* 183: 103–116.

HILLARP, N. A., K. FUXE, and A. DAHLSTROM, 1966. Demonstration and mapping of central neurons containing dopamine, norepinephrine and 5-hydroxytryptamine and their reactions to psychopharmaca. *Pharmacol. Rev.* 18:727–742.

HOLLISTER, L. E., 1962. Drug-induced psychoses and schizophrenic reactions—a critical comparison. *Ann. N.Y. Acad. Sci.* 96:80–88.

HOPE, J. M., E. CALLAWAY, and S. L. SANDS, 1951. Intravenous pervitin and the psychopathology of schizophrenia. *Dis. Nerv. Syst.* 12:67–72.

HORN, A. S., and S. H. SNYDER, 1971. Chlorpromazine and dopamine: Conformational similarities that correlate with the antischizophrenic activity of phenothiazine drugs. *Proc. Natl. Acad. Sci. USA* 68:2325–2328.

IVERSEN, L. L., 1967. *The Uptake and Storage of Noradrenaline in Sympathetic Nerves.* New York: Cambridge University Press.

IVERSEN, L. L., B. JARROTT, and M. A. SIMMONDS, 1971. Differences in the uptake, storage and metabolism of (±)- and (−)-norepinephrine. *Brit. J. Pharmacol.* 43:845–855.

JANOWSKY, D. S., M. K. EL-YOUSEL, J. M. DAVIS, and H. J. SEKERKE, 1973. Provocation of schizophrenic symptoms by intravenous methylphenidate. *Arch. Gen. Psychiat.* 28:185–191.

KALLMAN, S. J., 1938. *Genetics of Schizophrenia.* New York: Augustin Inc.

KETY, S. S., 1959. Biochemical theories of schizophrenia. A two-part critical review of current theories and the evidence used to support them. *Science* 125:1528–1532, 1590–1596.

KETY, S. S., 1972. Toward hypotheses for a biochemical component in the vulnerability to schizophrenia. *Sem. Psychiat.* 4: 233–238.

KLEIN, D. F., and J. M. DAVIS, 1969. *Diagnosis and Drug Treatment of Psychiatric Disorder.* Baltimore: Williams & Wilkins, pp. 52–138.

LEVINE, J., M. RINKEL, and M. GREENBLATT, 1948. Psychological and Physiological effects of intravenous Pervertin (methedrine). *Amer. J. Psychiat.* 105:429.

MAYER-GROSS, W., E. SLATER, and M. ROTH, 1960. *Clinical Psychiatry.* Baltimore: Williams & Wilkins, p. 377.

McDOWELL, F. H., 1970. Psychiatric aspects of L-dopa treatment in Parkinson's disease. In *L-Dopa and Parkinsonism,* A. Barbeau and F. McDowell, eds. Philadelphia: Davis, p. 321.

NYBACK, H., 1971. Effects of neuroleptic drugs on brain catecholamine neurons. *M.D. Thesis,* Stockholm: Kihlström and Söner Boktyckeri AB.

NYBACK, H., Z. BORZECKI, and G. SEDVALL, 1968. Accumulation and disappearance of catecholamines formed from tyrosine-C^{14} in mouse brain; effect of some psychotropic drugs. *Europ. J. Pharmacol.* 4:395–402.

PENNES, H. H., 1954. Clinical reactions of schizophrenics to sodium amytal, pervitin hydrochloride, mescaline sulfate and d-lysergic acid diethylamide. *J. Nerv. Ment. Dis.* 119:95–112.

PRINTZMETAL, M., and G. A. ALLES, 1940. The central nervous system stimulant effects of dextro-amphetamine sulphate. *J. Amer. Med. Assn.* 200:665–673.

SCHEEL-KRUGER, J., 1972. Behavioral and biochemical comparison of amphetamine derivatives, cocaine, benztropine and tricyclic antidepressant drugs. *Europ. J. Pharmacol.* 18: 63–73.

SNYDER, S. H., 1970. Putative neurotransmitters in the brain: Selective neuronal uptake, subcellular localization and interactions with centrally acting drugs. *Biol. Psychiat.* 2:367–389.

SNYDER, S. H., M. J. KUHAR, A. I. GREEN, J. T. COYLE, and E. G. SHASKAN, 1970a. Uptake and subcellular localization of neurotransmitters in the brain. *Intern. Rev. Neurobiol.* 13: 127–158.

SNYDER, S. H., K. M. TAYLOR, J. T. COYLE, and J. L. MEYERHOFF, 1970b. The role of brain dopamine in behavioral regulation and the actions of psychotropic drugs. *Amer. J. Psychiat.* 127:117–125.

SNYDER, S. H., and J. L. MEYERHOFF, 1973. How amphetamine acts in minimal brain dysfunction. *Ann. N.Y. Acad. Sci.* 205: 310–320.

SVENSSON, T. H., 1971. Functional and biochemical effects of d- and l-amphetamine on central catecholamine neurons. *Arch. Pharmakol.* 271:170–180.

TAYLOR, K. M., and S. H. SNYDER, 1970. Amphetamine: Differentiation by d- and l-isomers of animal behavioral effects involving central norepinephrine or dopamine. *Science* 168:1487–1489.

TAYLOR, K. M., and S. H. SNYDER, 1971. Differential effects of d- and l-amphetamine on behavior and on catecholamine disposition in dopamine and norepinephrine-containing neurons of rat brain. *Brain Res.* 28:295–309.

THOENEN, H., 1972. Chemical sympathectomy: A new tool in the investigation of the physiology and pharmacology of peripheral and central adrenergic neurons. In *Perspectives in Neuropharmacology,* S. H. Snyder, ed. New York: Oxford University Press, pp. 301–338.

UNGERSTEDT, U., 1971. Stereotaxic mapping of the monoamine pathways in the rat brain. *Acta Physiol. Scand.,* (suppl. 10) 367:1–48.

YARYURA-TOBIAS, J., B. DIAMOND, and S. MERLIS, 1970. The action of L-dopa on schizophrenic patients. *Curr. Ther. Res.* 12:528.

54

Schizophrenia:

Relationships to Dopamine Transmission,

Motor Control, and Feature Extraction

STEVEN MATTHYSSE

ABSTRACT Theories of tranquilizer action are reviewed, especially the hypothesis that tranquilizers act by blocking dopamine transmission; analogies are suggested between thought and affect in schizophrenia and the neural processes underlying motor control and feature extraction.

THERE IS COMPELLING evidence, as reviewed by Snyder in this Third Study Program, that phenothiazine tranquilizers are not merely sedatives but act on the core symptoms of schizophrenia. It used to be admitted that tranquilizing drugs were of value in getting patients to talk to their psychotherapists, but Philip May was probably closer to the truth when he suggested that psychotherapy with schizophrenic patients is of value insofar as it gets them to take their pills (May, 1968). The trouble with the phenothiazines from a scientific point of view is that they have a notoriously wide spectrum of actions: They are powerful electron donors (Foster and Fyfe, 1966); they inhibit oxidative phosphorylation (Medina, 1964); they stabilize cell membranes against lysis in distilled water (Despopoulos, 1970; Seeman et al., 1963). These effects, however, are not specific for the antipsychotic members of the phenothiazine group.

So many of the effects of phenothiazines are nonspecific, that Seeman has proposed the theory that "tranquilization is a type of 'selective regional anesthesia'," phenothiazines distinguishing themselves solely by tissue distribution (Seeman, 1966). Supporting evidence, as far as I know, has been obtained in only one case: Thiethylperazine, an antiemetic, concentrates in the cerebellum, whereas chlorpromazine, an antipsychotic, concentrates in subcortical areas (DeJaramillo et al., 1963). These concentrations were measured after only 90 min; it would be important to compare the regional brain distribution of chlorpromazine with a number of nonantipsychotic phenothiazines, and after long-term treatment.

The phenomenon of increased dopamine turnover after administration of tranquilizing drugs (interpreted as resulting from postsynaptic blockade of dopamine receptors) does not appear to be shared by the nonantipsychotic promethazine (Carlsson et al., 1963; Andén et al., 1964; Nybäck et al., 1968). This distinction suggests that dopamine blockade may be related to the *antipsychotic actions* of the phenothiazines, unlike the biochemical effects mentioned before. On the other hand, Parkinsonian motor side effects are prominent with antipsychotic drugs; because of the importance of dopamine in the extrapyramidal system, it is possible that the blocking action of tranquilizers on dopamine synapses could be related to their motor, rather than to their mental, effects. A drug of critical significance in separating mental from motor effects is thioridazine, a phenothiazine antipsychotic with only 3% incidence of Parkinsonian side effects (whereas fluphenazine, for example, has an incidence of 36%) (Cole et al., 1966). Single doses of thioridazine increased dopamine turnover (as indicated by caudate homovanillic acid) in the cat, but long-term treatment did not, although chlorpromazine continued to cause an increase under these conditions (Laverty et al., 1965). In mice, however, the two drugs had identical actions (O'Keeffe et al., 1970), so the implication of these findings is unclear.

Work in our laboratory is designed to investigate the specificity of the dopamine turnover effect for antipsychotic, as opposed to nonantipsychotic, phenothiazines. If a predicted similarity or difference fails to occur in an experimental animal, it could be argued that drug metabolism and tissue distribution differ between the animal and man; therefore, we felt it was desirable to use primates. Our method is to withdraw cerebrospinal fluid continuously from the lateral ventricle, one drop every 4 min for a month at a time, using each animal as his own control. In preliminary experiments, five nonantipsychotic phenothiazines had a negligible effect on homovanillic acid at the substantial dose of 10 mg/kg (diethazine,

STEVEN MATTHYSSE   Psychiatric Research Laboratories, Massachusetts General Hospital, Boston, Massachusetts fenethazine, methdilazine, promethazine, and pyrathiazine). In contrast, chlorpromazine typically causes a rise of 90% at this dose, and haloperidol and pimozide (both are antipsychotic) also cause large increases.

In these preliminary experiments, two drugs had effects on ventricular fluid homovanillic acid that do not seem entirely compatible with the dopamine hypothesis. Thiethylperazine, a nonantipsychotic phenothiazine, caused a rise of 48% at 0.3 mg/kg. The effect of thioridazine is disappointingly small: 8.5% at 10 mg/kg, in a typical experiment. We plan several refinements in our studies, which should clarify the actions of these two drugs: (1) detailed dose-response curves; (2) long-term administration of the drugs, an experiment for which the ventricular tapping procedure is especially suitable; (3) regional studies (it may be, for example, that thioridazine has a larger effect in the dopamine-containing nuclei of the limbic system than it does in the caudate, which is the structure most strongly represented in the ventricular fluid).

Holding in abeyance our remaining doubts, let us consider the implications of the hypothesis that the antipsychotic actions of the major tranquilizers (phenothiazines, butyrophenones, and diphenylbutylpiperidines) are brought about by blockade of dopamine receptors. We can begin by reflecting on the function of dopamine in the basal ganglia. The older view of the basal ganglia was that the motor cortex initiates movement, which is then modulated and bounded by the basal ganglia. The new view, presented at the Third Study Program by Kornhüber and by DeLong, is practically the reverse: Sensory input, processed in the whole cortex, results in "motor commands" transformed into action by the basal ganglia; once initiated in the basal ganglia, the action is then modulated and subjected to sensory feedback control by the motor area.

I think it is reasonable to propose that the nigrostriatal dopamine system regulates the *responsiveness* of the basal ganglia to these cortical commands, that is, the threshold for their emergence into real action. The strongest evidence is behavioral: In dopamine deficiency syndromes, such as Parkinson's disease, reserpine treatment or 6-hydroxydopamine lesion of the nigrostriatal dopamine tract (see Ungerstedt, this volume; S. Iversen, this volume), there is poverty of initiation of movement. Conversely, L-dopa accentuates choreiform movements and is proposed for detection of presymptomatic chorea by making latent movements manifest (Klawans et al., 1970; Cawein and Turney, 1971). Apomorphine, a dopamine receptor stimulator, causes stereotyped movements in animals (Ernst, 1965; Ungerstedt, this volume; S. Iversen, this volume). Neurons were observed in the caudate nucleus that were hyperpolarized and unresponsive to cortical stimulation, but these neurons underwent spontaneous depolarizing shifts of 5 mV lasting several minutes, during which they did respond to cortical stimulation. High-frequency nigral stimulation produced similar depolarizing shifts in caudate neurons lasting about 20 sec, during which their firing rate increased.

To summarize: Subthreshold motor commands are transformed into effectively initiated movements by the basal ganglia, and dopamine acts by disinhibition to permit the emergence of these movements from their subthreshold state. Now let us take a theoretical leap and suppose that dopamine neurons have the same function in the mental sphere. Because of the conservatism of evolution, a mechanism that was successful in one aspect of nervous function might be adopted in a similar, higher function. That is, analogy may be a way of nature as well as a way of thought.

Now if you introspect as you read this essay, you will observe that thoughts and images are continually rising near to the periphery of consciousness but fail to become fully emergent; some of these thoughts are cares of the day, some are dreamlike, disconnected images, some are even bizarre and disturbing. Imagine what would happen to an individual who did not have the capacity to keep these thoughts from taking over his field of awareness. I suggest that these preconscious thoughts and fantasies are like subthreshold motor commands; that some structure, analogous to the basal ganglia, releases them to enter the stream of consciousness; that dopamine neurons regulate the threshold for emergence of ideas, as they do the emergence of actions; that this disinhibitory system is overactive in schizophrenia; and that neuroleptic drugs increase "repression" of distracting ideas by blocking dopamine synapses.

The analogy may be illustrated by comparing two classical descriptions: Jung and Hassler's, of the function of the basal ganglia, and Eugen Bleuler's, of the thought process in schizophrenia. According to Jung and Hassler the basal ganglia contribute to motor control by "selectively inhibiting impulses from other sensory systems or from areas of activity which do not belong to the pattern of excitation most significant at the moment" (Jung and Hassler, 1959). Notice the similarity to Bleuler's conception of the missing controls in schizophrenic thinking: "It may be assumed that a certain force is necessary to keep associations in the track laid out by experience. Now it is possible that this force or 'control-tension' has also been diminished or hampered in its action because of the fundamental schizophrenic process" (Bleuler, 1924, pp. 81–82). Schizophrenics "are brought to a topic, totally irrelevant to the subject in hand, by any accidental things that happen to affect their senses . . . . The normal directives through questions from without and purposive conceptions from within are incapable of holding the train of thought·in the proper channels" (Bleuler, 1924, pp. 376–377).

So far we have an analogy between excessive emergence of subliminal ideas in schizophrenia, and excessive performance of subthreshold motor commands in certain diseases of the basal ganglia. However, schizophrenia is a disease of *affect* as well as of *thought*. To quote again from Bleuler: "One of the surest signs of the disease is the incapacity to modulate the affects, or an affective rigidity . . . the affective expressions are usually somewhat unnatural, exaggerated or theatrical. Consequently the joy of a schizophrenic does not transport us, and his expressions of pain leave us cold" (Bleuler, 1924, pp. 379–380).

The terms Bleuler uses are remarkable: *Incapacity to modulate, rigidity, exaggerated or theatrical.* Were he a neurologist describing disorders of the basal ganglia, these words would not be inappropriate to characterize chorea, athetosis, dystonia, and other hyper- and hypo-kinesias. Indeed, when affective rigidity and exaggeration become sufficiently intense, actual disturbances in posture and locomotion occur (catatonia). The affective disturbance in schizophrenia would suggest, to use Kornhüber's concept (1971; this volume), a disturbance in ramp or function generation, not in the basal ganglia, but in whatever system controls the initiation of affective response.

Where shall we look for a system, analogous to the basal ganglia, its threshold of activity regulated by dopamine neurons but more likely to be related to emergence of thought and to "ramp generation" of affect? Some fascinating observations have recently been made on the nucleus accumbens septi (Wilson, 1972). Embryologically and histologically, this nucleus is an extension of the head of the caudate, and like the caudate it has a major dopamine innervation (Ungerstedt, 1971). Its inputs, however, are from widespread parts of the limbic system, especially the hippocampus and pyriform cortex, just as the inputs to caudate are from widespread parts of the neocortex; indeed these inputs appear to be mutually exclusive. Its output is largely to the region of the pallidum known as substantia innominata. Its further projections are unknown, but it is interesting that DeLong has observed a response of substantia innominata neurons to the reward stimulus rather than solely to the motor response executed to earn the reward, suggesting that this region is involved in something more than motor control (DeLong, 1972).

The theory presented so far fails to account for certain features of schizophrenia.

(1) Identical twins of schizophrenics are not all schizophrenic (Pollin, 1972). The absence of complete concordance is thought to indicate an influence of environ-

ment. In the basal ganglia, on the other hand, one is struck by the lack of plasticity. The dyskinesias do not seem to be produced by, or improved by, experience.

(2) Interpersonal aversiveness seems to be a personality characteristic common to all forms of schizophrenia (Meehl, 1972).

(3) Schizophrenics have a strong tendency to misperceive social cues (Ploog, 1972).

A concept that has been a central focus of this Third Study Program goes far to fill these gaps: *Feature extraction.* It is plausible to assume that, in addition to cells that abstract from complex stimuli patterns, motion, and Fourier components, there should be cells that abstract *emotional significance* of complex stimuli for the organism. I will refer to these as *value detectors* because attractiveness-aversiveness, or value, a simple one-dimensional continuum, expresses the most universal emotional significance of stimuli; it remains to be seen whether there are also feature extractors specific for particular emotional meanings, such as danger or affection.

There is some evidence for the existence of value detectors. Neurons were observed in the pyriform cortex, hippocampus, and amygdala that changed in firing rate when a visual stimulus was presented that had, through conditioning, been associated with the opportunity of acquiring food by a lever press (Fuster and Ueyda, 1971). These units did not respond to the visual stimulus without the pairing, nor to a similar stimulus that had been associated with an occasion for avoiding shock. It would also be important to show that other occasions for reward, paired with the stimulus, could cause the stimulus to change firing rate in the same direction and to test the range of effective cue stimuli.

This addition to the theory does help account for the coexistence of genetic and environmental determinants in schizophrenia. An inborn skewness of the population of value detectors could cause interpersonal aversiveness and misperception of social cues. The work of Colin Blakemore (this volume) has shown that the population of visual feature detectors for angle can be skewed by exposure to horizontal or vertical stripes during a critical period. Perhaps exposure to continuous aversive stimuli in early life could similarly bias the population of value detectors in the direction of aversion rather than attraction. Blakemore also discovered cells that developed *bimodal* selectivities after alternating exposure to horizontal and vertical stripes during a critical period. It could be that erratic or illogical reward and punishment (Lidz et al., 1958), or simultaneous reward in one channel and punishment in another, as in Bateson's "double-bind" theory (1956), could cause value detectors to develop bimodal selectivities, and therefore lead to confused perception of social cues, and to interpersonal ambivalence.

This theory suggests enough experiments to keep us busy for a long time.

(1) Since there is more evidence for dopamine action on the "ramp generators" than on the feature detectors, tranquilizers should be more effective against thought disorder and affective rigidity than against interpersonal aversiveness and misperception of social cues.

(2) The theory suggests the importance of concentrated study of the connections, functions, electrophysiology, and pharmacology of the nucleus accumbens and other limbic dopamine nuclei, and of the actions of the mesolimbic dopamine system on them: In particular, does the dopamine system regulate the excitability of the nucleus accumbens?

(3) One ought also to investigate the existence of value detectors, using the complete set of criteria mentioned before. How widespread are they? Can the population be skewed by early experience? Does a drug like amphetamine, which can cause paranoid states (Snyder, 1972), modify their responsiveness? Do any of the diffuse ascending systems, such as those utilizing dopamine or norepinephrine, exert a biasing action (hyperpolarizing or depolarizing) on classes of value detectors? Are there "on" and "off" value detectors, by analogy with the visual system? An "off" value detector might be a *"sigh of relief" neuron*; an "on" value detector might be an *"I'm getting into trouble" neuron*. The latter type of feature extractor may be especially active in those who postulate theories of schizophrenia!

REFERENCES

ANDÉN, N. E., B. E. ROOS, and B. WERDINIUS, 1964. The effects of chlorpromazine, haloperidol and reserpine on the levels of phenolic acids in rabbit corpus striatum. *Life Sci.* 3: 149–158.

BATESON, G., D. D. JACKSON, J. HALEY, and J. H. WEAKLAND, 1956. Toward a theory of schizophrenia. *Behav. Sci.* 1: 251–264.

BLEULER, E., 1924. *Textbook of Psychiatry.* New York: Macmillan.

CARLSSON, A., and M. LINDQVIST, 1963. Effect of chlorpromazine or haloperidol on formation of 3-methoxytyramine and normetanephrine in mouse brain. *Acta Pharmacol. Toxicol.* 20:140–144.

CAWEIN, M., and F. TURNEY, 1971. Test for incipient Huntington's chorea. *New Eng. J. Med.* 284:504.

COLE, J. O., and D. J. CLYDE, 1966. Extrapyramidal side effects and clinical responses to the phenothiazines. *Rev. Canad. Biol.* 20:565–574.

DEJARAMILLO, G. A. V., and P. S. GUTH, 1963. A study of the localization of phenothiazines in dog brain. *Biochem. Pharmacol.* 12:525–532.

DELONG, M., 1972. Activity of basal ganglia neurons during movement. *Brain Res.* 40:127–135.

DESPOPOULOS, A., 1970. Antihemolytic actions of tricyclic tranquilizers. Structural correlations. *Biochem. Pharmacol.* 19: 2907–2914.

ERNST, A. M., 1965. Relation between the action of dopamine and apomorphine and their O-methylated derivatives upon the CNS. *Psychopharmacologia* 7:391–399.

FOSTER, R., and C. A. FYFE, 1966. Electron-donor-acceptor complex formation by compounds of biological interest. II. The association constants of various 4-dinitrobenzene-phenothiazine drug complexes. *Biochim. Biophys. Acta.* 112: 490–495.

FUSTER, J. M., and A. A. UEYDA, 1971. Reactivity of limbic neurons of the monkey to appetitive and aversive signals. *Electroenceph. Clin. Neurophysiol.* 30:281–293.

JUNG, R., and R. HASSLER, 1959. The extrapyramidal motor system. In *Handbook of Physiology*, J. Field, ed. Baltimore: Williams & Wilkins, Vol. 2, pp. 863–927.

KLAWANS, H. C., G. W. PAULSON, and A. BARBEAU, 1970. Predictive test for Huntington's chorea. *Lancet* 2:1185–1186.

KORNHÜBER, H. H., 1971. Motor functions of cerebellum and basal ganglia: The cerebellocortical saccadic (ballistic) clock, the cerebellonuclear hold regulator, and the basal ganglia ramp (voluntary speed smooth movement) generator. *Kybernetik* 8:157–162.

LAVERTY, R., and D. F. SHARMAN, 1965. Modification by drugs of the metabolism of 3,4-dihydroxyphenylethylamine, noradrenaline and 5-hydroxytryptamine in the brain. *Brit. J. Pharmacol.* 24:759–772.

LIDZ, T., A. CORNELISON, D. TERRY, and S. FLECK, 1958. Intrafamilial environment of the schizophrenic patient: VI. The transmission of irrationality. *Arch. Neurol. Psychiat.* 79: 305–316.

MAY, P. R. A., 1968. Anti-psychotic drugs and other forms of therapy. In *Psychopharmacology: A Review of Progress 1957–1967*, D. H. Efron, ed. Washington, D.C.: Public Health Service, pp. 1155–1176.

MEDINA, H., 1964. The effect of certain phenothiazines on the structure and metabolic activity of sarcosomes of the guinea pig heart. *Biochem. Pharmacol.* 13:461–467.

MEEHL, P. E., 1972. In *Prospects for Research on Schizophrenia*, S. S. Kety and S. Matthysse, eds. *Neurosciences Res. Prog. Bull.* 10:377–380.

NYBÄCK, H., Z. BORZECKI, and G. SEDVALL, 1968. Accumulation and disappearance of catecholamines formed from tyrosine-^{14}C in mouse brain; effect of some psychotropic drugs. *Euro. J. Pharmacol.* 4:395–403.

O'KEEFFE, R., D. F. SHARMAN, and M. VOGT, 1970. Effect of drugs used in psychoses on cerebral dopamine metabolism. *Brit. J. Pharmacol.* 38:287–304.

PLOOG, D., 1972. In *Prospects for Research on Schizophrenia*, S. S. Kety and S. Matthysse, eds. *Neurosciences Res. Prog. Bull.* 10: 394–396.

POLLIN, W., 1972. The pathogenesis of schizophrenia: Possible relationships between genetic, biochemical, and experiential factors. *Arch. Gen. Psychiat.* 27:29–37.

SEEMAN, P. M., 1966. Membrane stabilization by drugs: Tranquilizers, steroids, and anesthetics. *Int. Rev. Neurobiol.* 9:145–211.

SEEMAN, P. M., and H. S. BAILY, 1963. The surface activity of tranquilizers. *Biochem. Pharmacol.* 12:1181–1191.

SNYDER, S. H., 1972. In *Prospects for Research on Schizophrenia,* S. S. Kety and S. Matthysse, eds. *Neurosciences Res. Prog. Bull.* 10:433–435.

UNGERSTEDT, U., 1971. Stereotaxic mapping of the monoamine pathways in the rat brain. *Acta Physiol. Scand.* (Suppl.) 367: 1–48.

WILSON, R. D., 1972. The neural associations of nucleus accumbens septi in the albino rat. Ph.D. Thesis, Dept. of Psychology, Mass. Inst. of Technology. (This work was done in association with Drs. L. Heimer and W. J. H. Nauta.)